The CRETACEOUS FOSSILS OF SOUTH-CENTRAL AFRICA: an illustrated guide

M. R. Cooper

CRC Press is an imprint of the
Taylor & Francis Group, an **informa** business
A BALKEMA BOOK

CRC Press/Balkema is an imprint of the Taylor & Francis Group, an informa business

Typeset by Apex CoVantage, LLC

Library of Congress Cataloging-in-Publication Data
Applied for

Published by: CRC Press/Balkema
Schipholweg 107c, 2316 XC Leiden,The Netherlands
e-mail: Pub.NL@taylorandfrancis.com www.crcpress.com – www.taylorandfrancis.com

ISBN: 978-1-138-33650-6 (Hbk)
ISBN: 978-0-429-44297-1 (eBook)

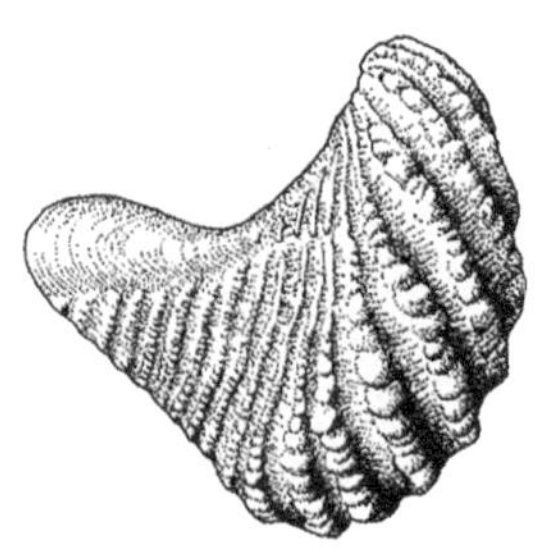

Cover design and illustrations by M. R. Cooper

Set in 9-11pt Times New Roman

Index

Introduction 1
Eustasy 1
Milankovitch cycles 2
The breakup of Gondwana 2
Vicariance 2
The terminal Cretaceous extinction event 3

REGIONAL GEOLOGY

Namibia 3
Northern Cape 4
Western Cape 5
Lesotho 5
Swaziland 5
Eastern Cape 5
KwaZulu-Natal 10
Moçambique 19
Zimbabwe 23
Angola 24
Malawi 29
Zambia 29
Tanzania 29
References 31
Illustrations of Cretaceous fossils 51

About the author

Michael Cooper is Honorary Curator of Palaeontology at the Durban Natural Science Museum and an Emeritus Professor of Geology at the University of KwaZulu-Natal. He has studied the Cretaceous geology and palaeontology of south-central Africa for over 40 years, concerned mainly with the biostratigraphy, evolution and classification of the molluscan faunas (particularly ammonite and bivalves). He has published over 120 scientific articles on these subjects, and his studies have involved fieldwork in South Africa, Zimbabwe, Mozambique and Angola.

Introduction

Deriving its name from the widespread chalk sequences of Western Europe, the Cretaceous has a marine connotation. This is hardly surprising since the period dates a transgressive maximum, when sea levels were some 250m above present. As a result marine deposits of Cretaceous age crop out along many parts of the continental margins of Southern and Central Africa, as well as building much of the continental shelves. Along the east coast they attain their fullest development and greatest surface extent on the South Moçambique-Maputaland coastal plain and, along the west coast, in the Kwanza Basin of Angola.

As elsewhere in the world, the Cretaceous history of the continental margins of Africa was one of alternating marine transgressions and regressions brought about by fluctuations in sea level. In the marine beds, ammonites are of prime biostratigraphic importance, their diversity, rapid evolution, often pelagic habits, buoyant shells and, hence, widespread distribution facilitating detailed inter-regional correlation often on a world-wide scale. In this respect however inoceramid bivalves are increasing in importance.

Generally, terrestrial deposits are present at the base of most Cretaceous marine sequences, as well as cropping out sporadically in the continental interior. They were also intercalated in marine sequences at times of lowered sea level (regression).

Eustasy

Eustasy is worldwide changes in sea level. These may be of tectonic or climatic origin. Sea-level rise results in flooding of continental margins and **marine transgression**; shorelines move inland and marine deposits overstep terrestrial. A drop in sea level results in shoreline retreat, **marine regression** and sea floor exposure; terrestrial deposits overlie marine. There are multiple causes for sea-level change, and it is possible to recognize 7 orders of cyclicity.

1st-order cyclicity is due to the build-up of mantle heat beneath supercontinents. It results from mantle turnover and is recorded by **megacycles** (= Wilson cycles) with a periodicity of about 320 Ma. They are documented by the volcanosedimentary fill of **cratonic basins**. In Southern Africa the Cretaceous is part of the **Kazu Megacycle** which commenced in the Late Triassic and continues to the present. The transgressive phase of this megacycle peaked in the Early Maastrichtian. The Late Maastrichtian and Cainozoic are part of a 1st-order regression, with sea levels falling intermittently but progressively.

2nd-order cycles have a periodicity of 75-80 Ma and are of plate-tectonic origin. These **supercycles** are of the same order as the creation and destruction of major mountain ranges. The Cretaceous was a period of 2nd-order transgression, the so-called **Great Cretaceous Transgression**. Sea level rose gradually but intermittent until its dramatic fall at the close of the Maastrichtian.

3rd-order cycles have a periodicity of 16-18 Ma. The Great Cretaceous Transgression involved 5 such **macrocycles**, with peak transgressions dating to the Late Valanginian/Early Hauterivian, Late Early Aptian, Late Albian (but not latest), Late Santonian/Early Campanian and Early Maastrichtian. Perhaps they are related to **intra-plate magmatism** since certain igneous events have periodicity of this order.

4th-order cycles have a periodicity of 2-5 Ma. These **epicycles** are a product of ocean-floor spreading. Episodic release of mantle heat at the mid-oceanic ridges produces volumetric changes in the ridge system and fluctuations in the rate of ocean-floor spreading. Thermal expansion raises sea level and results in marine transgression. In a number of cases 4th-order regressions, due to thermal contraction of the ridges, were accompanied by onshore volcanism.

Milankovitch cycles

The rotation of the earth around the sun is responsible for fluctuations in the amount of heat received by the earth. This orbital forcing of climates is responsible for **Milankovitch cycles** of different periodicity.

Since the earth's orbit is elliptical, not circular, gravitational pull is greatest when the sun is closest to the earth and least when it is furtherest away. This **orbital eccentricity** is responsible for **5th-order cycles** with periodicities of 405 and 100 ka. These are documented by cyclical rhythms in sedimentation (**cyclothems**).

6th-order cycles with a periodicity of 30 ka result from changes in the **axis of rotation** which vary between 21.5-24.5° to the plane of the earth. These **microcycles** are recorded in tidally-influenced sedimentary deposits (tidalites).

7th-order cycles with a periodicity of 23 ka result from the **precession** of the equinoxes produced by a slight wobble in the earth's axis. This changes the time of the solstices (longest and shortest days) and results in abnormally warm or cold summers. These **solsticycles** have their greatest effect in periglacial regions but, since the Cretaceous had ice-free poles, are unlikely to be recognized in this period.

The breakup of Gondwana

Although fragmentation of Gondwana was largely a Jurassic event, separation of Africa from South America occurred only during the Cretaceous. Fossiliferous Late Jurassic deposits are present on the eastern Falkland Plateau which, at that time, was adjacent to KwaZulu-Natal. In addition onshore Middle Jurassic (Bathonian) marine sediments are well known from Madagascar and latest Jurassic (Tithonian) strata crop out in northern Moçambique. These exposures indicate the existence of a narrow seaway the length of Africa prior to the Cretaceous.

The west coast separation of South America was a later event than the east coast split. Although rifting commenced already in the Late Jurassic, ocean-floor spreading of the South Atlantic is dated to the Valanginian (c.139 Ma). Final separation of the South American and African plates seems to have been an Albian event. Nonetheless by Neocomian time the margins of Southern and Central Africa were much as they are today, and widespread marine sedimentation had commenced on the newly formed continental shelves.

Vicariance

The continental distribution of Late Jurassic - Early Cretaceous organisms was disrupted by the breakup of Gondwana. The consequent physical barriers to gene flow, i.e. the formation of the Atlantic and Indian Oceans, resulted in **disjunct distributions** and produced **vicariance**. This provides evidence of ancient faunal compositions even in the absence of fossils.

Distribution of the extinct Early Permian pteridosperm *Glossopteris* on all southern continents provided compelling evidence for the theory of continental drift. In 1927 K. H. Barnard referred to the usefulness of this theory in explaining the distribution of living freshwater isopods and amphipods.

Among the Orthoptera, the camel crickets (family Rhaphidophoridae) display a classic Gondwanic distribution, as do certain hymenopterans (Scolebythidae) and earthworms (Oligochaeta) of the subfamilies Alminae and Acanthodrillinae. Freshwater fishes also show vicariant distributions, with Cichlidae, Odontoglossidae and Characinidae present already in the Early Cretaceous rivers of Gondwana. Among birds the Struthioniformes, including living rhea (South America), ostrich (Africa), cassowary (Australia) and emu (New Zealand), display a classic Gondwanic distribution pattern which points to an origin in the Early Cretaceous.

The terminal Cretaceous extinction event

Fluctuations in sea level control the earth's climates and life on earth.The spread of shallow epicontinental seas has a moderating effect, with the extension of maritime influences, reduction in the elevation of landmasses, increased evaporation and precipitation, and general climatic warming. Flooding of the continental margins introduces nutrients into the oceans, stimulating phytoplankton proliferation and increased liberation of photosynthetic O_2. Conversely drops in sea level result in the spread of continental climates with increased elevation of landmasses, greater thermal range, reduced rainfall, elevated CO_2 and the spread of aridity. These environmental shifts impact all life on earth.

The Late Maastrichtrian was a time of profound regression which saw sea levels drop below the continental break, exposing the continental shelves. Since the greatest diversity of life in the oceans occurs on the continental shelves, obliteration of this environment had a devastating effect on marine life. This, and the induced climatic and environmental changes, were the main causes of end-Cretaceous extinctions, exacerbated by a major phase of volcanism in India (the Deccan Traps), kimberlite magmatism in Africa and, coincidently, a large meteorite impact in the Gulf of Mexico. Claims that the latter was the prime cause of the end-Cretaceous mass extinction are mere Disneyan media hype.

Namibia

Most onshore Cretaceous exposures in Namibia are terrestrial and volcanic. The volcanics are linked genetically to opening of the South Atlantic, with the bulk of the extrusives preserved in the Paraná basin of eastern South America. Scattered exposures of torrential fluviatile clastics and capping pedocretes, assigned to the **Dasdab Formation**, occur along the margin of the Bushmanland Plateau and Namaqua Escarpment. Although without fossils they are believed to be Cretaceous. A Barremian to Maastrichtian sequence of marine rocks exists on the continental shelf off southern Namibia, but only a single small outlier occurs on land.

Koako Group

The Koako Group is a bimodal succession of terrestrial volcanics well represented in the southern Koakoveld. The sequence is divided into:

ii) **Grootberg Formation** - up to 150 m of hard pale yellowish-grey rhyolites and orthoclase porphyries with a sharp basal contact.

i) **Etendeka Formation** - 450 m of massive, blue-grey amygdaloidal basalts with occasional sedimentary interbeds.

The Etendeka basalts are olivine-rich tholeiites, with two flows of porphyritic tachylyte north of Gai-as. Rapid weathering of the basalts produces a stepped topography. A radiometric age of 138-128 Ma for Paraná volcanics indicates a Late Valanginian-Late

Hauterivian age for the Etendeka. Thin interbeds of aeolian sandstone up to 200 m above the base testify to persistence of the **Stormberg Desert**.

The resistance to weathering of the Grootberg porphyries is responsible for the mesa-and-butte topography of the region.

Wanderveld IV Formation

Strongly transgressive Cretaceous strata rest with marked unconformity on Precambrian dolomites of the Bogenfels Formation in southern Namibia. Near Bogenfels the following section is recorded:

iv.) About 2 m of yellowish-weathering olive-green silts.

iii.) **Oyster shell bed** - a hard, calcified *Rhynchostreon* lumachelle about 30 cm thick.

ii.) 1.8 m of yellowish silty sand with abundant oysters and heterodont bivalves.

i.) About 2 m of grit and matrix-supported conglomerate.

Abundance of the oyster *Rhynchostreon suborbiculatum*, together with the ammonite *Proplacenticeras memoriaeschloenbachi* (= *P. merenskyi*), suggest a Late Cenomanian age for the formation. However the microfaunas are dated to the Santonian and more work is required.

Northern Cape

Slow subsidence of the Kalahari basin commenced soon after break-up of Gondwana and continues to the present. Well removed from the shorelines of the time, basin fill was of terrestrial sediment.

Kalahari Group

A Cretaceous age for commencement of Kalahari sedimentation is indicated by the discovery of a dinosaur on the farm Kangnas, 15 km south of Goodhouse. Here some 3 m of coarse colluvium, with calcrete and silcrete nodules, are developed at the base, yielding silicified wood and bones of the ornithischian dinosaur *Kangnasaurus*. The precise age is not known.

Deeply-incised palaeo-valleys in the Knersvlakte represent the drainage system of the **Karoo River**. Sediment fill is lacking but fission-track dating suggests an Aptian age (120-110 Ma) for this erosive event. Both Early and Late Cretaceous alluvial gravels are known from **Mahura Muthla**, 60 km east of Kuruman, and include fossil wood (*Podocarpoxylon umzambense*).

In the NW Cape, centered around the hamlet of Gamoep, is a cluster of more than 270 kimberlite pipes and diatremes, mostly buried beneath Kalahari sand. Those that have been dated bracket between 71.6-64.2 Ma (Late Campanian - earliest Palaeocene). Due to the rapid decomposition of kimberlite many pipes are capped by fossiliferous paludal deposits which, in the **Arnot Pipe**, have yielded the pipid frog *Eoxenopoides*, fossil wood, leaf fragments and plant spores (palynomorphs).

Further to the east, around Kimberley, another cluster of kimberlites fall into two groups on the basis of composition and age. The older **Group 1 kimberlites** are more numerous and date between 140-114 Ma (Valanginian-Late Aptian), whereas the younger **Group 2 pipes** cluster around 90 Ma (Late Turonian), with the ages of the 5 major pipes between 100-75Ma (Cenomanian-Campanian).

WESTERN CAPE

Cretaceous sedimentary rocks are unknown from the Western Cape. However evidence of magmatic activity is provided by the **Cape Peninsula dyke swarm**, a group of NW-trending dolerite dykes with an age of 132 Ma (Early Hauterivian). Emplacement therefore was coeval with Paraná-Etendeka volcanism, and relates to opening of the South Atlantic.

LESOTHO

Lesotho has the highest concentration of kimberlite pipes and dykes in the world. Dating of the **Mothae kimberlite** to 87 Ma indicates a Late Cretaceous (Late Coniacian) age for some.

SWAZILAND

No rocks of Cretaceous age have been reported from Swaziland, the only dated kimberlite being Permian.

EASTERN CAPE

Late Jurassic rift tectonics initiated development of the **Algoa Basin**, and the basin fill is assigned to the **Uitenhage Group**. Deposition commenced in fault-bound troughs, mostly half-grabens, and initially was of terrestrial Late Jurassic conglomerates of the **Enon Formation**. With continued subsidence the Algoa Basin accumulated a thick succession of fluviatile and shallow-marine sediments of earliest Cretaceous age.

Kirkwood Formation

As a matter of convenience the Jurassic/Cretaceous boundary in the Algoa basin is drawn at the base of the Kirkwood Formation, a Berriasian to Middle Valanginian unit of predominantly fluviatile siliciclastics. Surface exposures are best developed at and around the town of Kirkwood where boreholes proved a maximum thickness of 2100 m and allow for a threefold subdivision:

iii.) **Bezuidenhouts Member** - 550 m of variegated shales and siltstones with subordinate upward-fining, cross-bedded sandstones and pebble washes.

ii.) **Colchester Member** - 350 m of grey to black marine shales, siltstones and sandstones with some thin evaporites.

i.) **Swartkops Member** - up to 63 m of poorly-sorted, fine- to medium-grained quartzitic sandstones with thin interbeds of grey shale.

Where Enon conglomerate is absent, Swartkops sandstones rest directly upon older rocks and, near Centlivres in the Coega valley, fill crevices in the underlying Ordovician sandstones.

The Colchester Shale is best known from the subsurface and at outcrop is represented only by a few thin marine intercalations in the Bezuidenhouts Member. Such exposures occur on the farm Dunbrodie, at North End Lake, and again at the Bethelsdorp Salt Pan. Although the age of the Colchester Member is uncertain, the presence of the Berriasian belemnite *Belemnopsis gladiator* in the Algoa basin is likely to date this unit. Other fossils from the Bethelsdorp Salt Pan include the bivalves *Arctica*, *Corbicula*, *Camptonectes*, *Lycettia* and *Isoperna*, as well as the gastropods *Actaeonina*, *Limnaea*, *Turritella*, *Turbo* and *Viviparus*, a mixture of freshwater and marine forms.

Of the three members of the Kirkwood most outcrops are of the Bezuidenhouts Member. This fluvio-lacustrine unit grades laterally and vertically into marine facies and comprises mainly mudstones and sandstones stacked in fining-upward fluviatile rhythms. Deposition was by meandering streams on a mature low-lying alluvial plain. Yellowish channel sandstones make up to 40% of the succession and pass distally into massive, intensely-bioturbated estuarine sandstones with frequent calcareous nodules and large fossil logs. Lags at the base of some of the sandstones contain bone fragments. Better-preserved vertebrate remains are found in the Bezuidenhouts River-Dunbrody area and include the dinosaurs *Algoasaurus*, *Palaeoscincus*, *Paranthodon* and *Nqwebasaurus*, as well as crocodile, tortoise, lizard, frog and fish remains. At Mfuleni, a thin lacustrine unit of grey limestones and shales is rich in freshwater ostracods (*Cypridea*, *Bisulcocypris*, *Darwinula*) and small gastropods (*Viviparus*, *Psammobia*).

Lakes and swamps were frequent on the overbank areas of the Bezuidenhouts floodplain. Ferns (*Sphenopteris*, *Osmundites*, *Onychiopsis*), cycads (*Palaeozamia*, *Otozamites*, *Pterophyllum*, *Dictyozamites*, *Nilssonia*) and bryophytes lined the watercourses, and large conifers with *Brachyphyllum* foliage and *Araucarites* cones occupied high ground.

Eastwards the Kirkwood Formation grades into the **Infanta Shale**, a poorly-known offshore marine unit reported only from the subsurface. It consists mostly of pale to dark grey shales, marls and siltstones.

Sundays River Formation

This marine formation follows conformably on the Kirkwood and, due to transgressive toplap, rests on various underlying units. The following subdivisions have been recognized in boreholes:

iv.) **Vetmaak Member** - 108 m of grey to greenish, very fine- to medium-grained lignitic sandstones alternating with fossiliferous grey mudstones.

iii.) **Addo Member** - 182 m of fossiliferous grey lignitic siltstones alternating with thin, grey, very fine- to medium-grained lignitic sandstones and dark grey thinly-bedded shales.

ii.) **Soetgenoeg Member** - 350 m of very fine- to coarse-grained lignitic and calcareous sandstones, typically in graded cycles, as well as fossiliferous grey lignitic shales. Convolute bedding, scour-and-fill and bioturbation structures are common.

i.) **Amsterdamhoek Member** - 357 m of bluish-grey to black fossiliferous shales with subordinate grey fine- to very fine-grained shelly and lignitic sandstones in fining-upward cycles.

At outcrop the Sundays River Formation comprises mainly grey to greenish shales, with immature sandstones making up to 40% of the pile. A few thin bentonites have been intersected in boreholes. The sandstones are like those of the Kirkwood but somewhat more mature, with abundant current ripples, trough and planar cross-bedding sometimes with bipolar foresets, occasional herringbone cross-stratification, clay galls, channel lags with mud intraclasts, and abundant plant and shell debris. In many places, however, primary structures were obliterated by intense bioturbation. Primary structures in the mudrocks include flaser and wavy bedding, locally with load casts and convolute bedding. Deposition was on tidal mud flats with the absence or extreme rarity of echinoderms, bryozoans, brachiopods, corals, planktonic formanifera, crustaceans and fishes suggesting a marginally-marine depositional environment. The presence of gypsum on bedding planes and joints indicates a tendency to hypersalinity.

Correlation between the borehole lithostratigraphy of the Sundays River Formation and surface exposures is difficult. Outcrops of the **Amsterdamhoek Member** occur in the Swartkops railway cutting, near Coega

Kop, along the Addo-Port Elizabeth road, and are worked in the Swartkops brick quarries where they have yielded a rich molluscan fauna dominated by the bivalves *Aetostreon*, *Megacucullaea*, *Herzogina*, *Gervillella*, *Rinetrigonia* and *Steinmanella*, together with *Camptonectes*, *Entolium*, *Thetis*, *Lima*, *Modiolus*, *Lycettia*, *Pholadomya*, *Myopholas?*, *Nucula* and many others. Gastropods include *Natica*, *Neritopsis*, *Patellona*, *Alaria* and *Bathrotomaria*. Rarer elements include the ammonites *Olcostephanus*, *Neohoploceras* and *Criosarasinella*, the nautiloid *Eutrephoceras*, corals (*Isastraea*, *Thamnastraea*), the ophiuroid *Ophiolancea*, the shrimp *Meyeria*, the sponge *Pachastrella* and the plesiosaur *Leptocleidus*. It is of early Late Valanginian age.

Prime exposures of the Soetgenoeg Member are encountered in the Sundays River valley, on the farm Zoetgeneugd, where they have yielded *Olcostephanus*. The Addo Member is restricted to a single section near Eb'-en-Vloed. The Vetmaak Member is believed to have the widest surface outcrop of all, and is exposed in the cliff section immediately west of Colchester, near the top of which is a storm accumulation, of *Rinetrigonia ventricosa*. Ammonites from this section, dated microfaunally to the Late Hauterivian, include *Bochianites* and *Phyllopachyceras*.

Robberg Formation

The Robberg Formation comprises up to 160 m of silicified conglomerates and sandstones resting with profound unconformity on Ordovician sandstones of the Table Mountain Group. Outcrops are restricted to the Robberg, a peninsula in Plettenberg Bay, and coastal exposures a few kilometres to the west.

At Cape St Blaize the base of the Robberg is marked by a thin sedimentary breccia, overlain by hard whitish sandstones with lenses of breccia, abundant mud intraclasts, some carbonate-rich layers, and fragments of fossil wood. At Guanogat, and near Rondeklippe, the basal Robberg is preserved in narrow steep-sided palaeo-valleys and comprises an unsorted to crudely stratified accumulation of terrestrial rockfall debris and slide deposits. Invariably these talus breccias are overlain with erosive contact by terrestrial colluvium. Some grey sandstone lenses near the base may be tuffaceous.

The bulk of the Robberg follows with abrupt contact on the colluvium and comprises large-pebble to cobble conglomerates with well-rounded clasts of Table Mountain sandstone. Beds are internally massive, up to 18 m thick, and sometimes fine upwards. Sorting is impressive and suggests deposition in a high-energy beach environment, with conspicuous imbrication pointing to flow from the north. Lenses of sandstone are common and locally conglomerates fill estuarine channels.

Robberg conglomerates pass vertically upwards and laterally into some 90 m of fossiliferous marine sandstones at the base of which is a channel facies of stacked cycles of quartz arenite up to 4 m thick. Trough cross-bedding is conspicuous, often with rippled foresets, as well as herringbone cross-stratification and *Skolithos* burrows. This facies grades upwards into laminated and bioturbated sandstones with mud-draped ripples and rare desiccation cracks. Trace fossils include horizontal burrows of *Rhizocorallium* and *Scalarituba*, asteroid resting traces, and trails resembling *Isopodichnus*. Deposition was under shallow-subtidal to upper-intertidal conditions, with evidence for several minor fluctuations in sea level.

Robberg fossils are mostly poorly-preserved internal moulds, including the bivalves *Iotrigonia*, *Gervillella* and *Megatrigonia*, an indeterminate ammonite, ribbed stems of *Neocalamites*, fern scraps and cycad fragments.

In the Knysna area, the so-called "Enon" derives from the north, with deposition mainly by braided streams. However, bedding is locally prominent and may reflect reworking in a nearshore shallow-marine environment. These factors suggest the "Enon"of this area may be an unsilicified facies of the Robberg.

Brenton Formation

Marine strata of the Brenton Formation are exposed in low muddy banks along the SW shore of Knysna lagoon, at Brenton-on-Lake. They comprise richly-fossiliferous sticky grey mudstones which coarsen upwards into imbricated beach conglomerates with the bivalves *Mayesella* and *Pterotrigonioides*, and a plesiosaur tooth. The mudstones contain a rich but fragile bivalve fauna dominated by *Neuquenella kensleyi*, with which are associated *Rinetrigonia*, *Ptychomya*, *Meretrix* and *Isognomon*, a nautiloid, a controversial ammonite fragment and the belemnite *Hibolites*. Although similar to those of the Sundays River Formation, the trigonias are different species and the Brenton Formation is of slightly different age, perhaps late Late Valanginian.

Since the Knysna Heads, prominences of Ordovician quartzite, were in existence at the time of Brenton sedimentation, the formation is a back-barrier muddy lagoonal deposit whose diverse fauna indicates stenohaline conditions.

Igoda Formation

This is a small outlier of Late Cretaceous strata some 15 km SW of East London. Here it rests unconformably on Late Permian sediments of the Beaufort Group and is subdivided into:

Member III - up to 20 m of glauconitic sandy limestones and calcareous sandstones with a shelly fauna dominated by the oysters *Rhynchostreon* and *Actinostreon*.

Member II - a thin (0.5 m) small-pebble conglomerate with a matrix of glauconitic sandy limestone.

Member I - an impersistent unit, up to 3 m thick, of poorly-consolidated unfossiliferous white sandstone with pebble lags.

The presence of the ammonites *Baculites subanceps*, *Pachydiscus* and *Saghalinites* indicate a Late Campanian age. Also present are bivalves (*Linotrigonia*, *Spondylus*), the gastropod *Zaria*, the brachiopod *Eolacazella* and the coral *Caryophyllia*.

Lower Need's Camp Formation

The occurrence of the brachiopod *Eolacazella affine* at Need's Camp is the basis for correlation with the Igoda Formation and a Late Campanian age, although the facies are different. At this site, on the road between East London and Kingwilliamstown, some 1.6 m of chalky bryozoan limestone is exposed resting unconformably on silcrete and dolerite at an altitude of 374 m. Macrofossils include the bivalves *"Inoceramus"*, *Isognomon*, *Rhynchostreon* and *Phygraea*, the regular echinoid *Coptosoma*, the brachiopod *Eolacazella*, the coral *Caryophyllia*, sharks' teeth (*Carcharias*, *Isurus*) and a plesiosaur tooth. Sedimentation was in a clear-water, stenohaline, back-barrier lagoon.

Mngazana Formation

The Mngazana is a tiny downfaulted outlier preserved around the base of Mqualeni Hill, at the Mngazana estuary in the Eastern Cape. Here it rests unconformably on Late Triassic sandstones of the Molteno

Formation and comprises intensely hard greenish-grey conglomerates alternating with grits and sandstones containing lenses of white-weathering carbonaceous and calcareous sandstone. Surf-zone deposition was on a shoreline alluvial cone backed by a fault scarp. Marine fossils include the ammonite *Umgazaniceras*, the bivalves *Steinmanella* and *Indogrammatodon*, a belemnite close to *Hibolites subfusiformis*, the cycads *Dictyozamites* and *Otozamites*, and the fern *Onychiopsis*.

Mbotyi Formation

Some 40 km north of Mngazana are 300 m of poorly-sorted conglomerates and sedimentary breccias of the Mbotyi Formation, with lenses of coarse-grained sandstone and grey-green mudstone. So far the formation has proven unfossiliferous, save for carbonized wood chips in some of the sandy lithosomes. The succession is downthrown to the north, against Palaeozoic Msikaba sandstones, by the **Egosa Fault**. Close lithological similarity and geographical proximity favour correlation with the Mngazana.

Mzamba Formation

The Mzamba is exposed intermittently along the littoral south of Port Edward. Here it rests with marked unconformity on Palaeozoic sandstones of the Natal Group, with a basal conglomerate of quartzarenite and lydianite pebbles. Large current-orientated *Teredo*-bored logs several metres in length are common, and a rich invertebrate fauna points to shallow-marine nearshore sedimentation. The fossil logs are mostly Monimiaceae (*Protoatherospermoxylon*, *Hedycaryoxylon*, *Paraphyllanthoxylon*, *Securinegoxylon*) but also with some Euphorbiaceae (*Bridelioxylon*) and are indicative of tropical to subtropical climates. Hence Late Cretaceous environments along the southeast coast of Africa were much as they are today.

The presence of the ammonites *Texanites*, *Plesiotexanites*, *Texasia* and *Baculites capensis*, as well as the bivalve *Cladoceramus undulatoplicatus*, near the base date the commencement of sedimentation to the Early Santonian. The overlying succession, which spans the remainder of the Santonian and the earliest Campanian, comprises storm-generated small-scale fining-upward sedimentary cycles (*tempestites*). Each cycle commences with an intensely-hard current sorted gritty shellbed overlain by greyish-green fine-grained siltstones capped by thin lenses and drapes of black bioturbated mudstone. Frequent minor breaks in the succession are recorded by serpulid- and oyster-encrusted hardgrounds with a prominent layer of hiatus concretions taken to mark the Santonian-Campanian boundary.

There is a very rich and diverse invertebrate fauna. Besides those already mentioned, Santonian elements include the ammonites *Texasia*, *Pseudoschloenbachia*, *Pseudophyllites*, *Damesites*, *Saghalinites*, *Hyporbulites*, *Gaudryceras*, *Gardeniceras*, *Natalites*, *Pseudoxybeloceras* and *Madagascarites*, the echinoids *Hemiaster*, *Cardiaster*, *Cassidulus* and *Pseudodiadema*, the gastropods *Actaeonella*, *Zaria*, *Spirocolpus*, *Confusiscala*, *Cancellaria*, *Paleopsephaea*, *Pyropsis*, *Afrocypraea*, *Deussenia*, *Microgaza*, *Solariella*, *Gymnaris* and many others. The bivalves include *Neithea*, *Acanthotrigonia*, *Linotrigonia*, *Trigonarca*, *Glycymeris*, *Nordensjkoeldia*, *Platyceramus*, *Trachycardium*, *Acanthocardia*, *Crassatellites*, *Trigonocallista*, *Veniella* and many more. Early Campanian elements of the fauna include *Submortoniceras*,

Pseudoschloenbachia griesbachi, *Eulophoceras*, *Neopachydiscus* and *Argentiscaphites*, with a shell-bed rich in the pectinoid *Camptonectes kaffraria* near the top of the section.

A temporary exposure of the Mzamba Formation was excavated at the Wild Coast Casino, its basal conglomerate 34 m above sea level. The diverse fauna included the ammonites *Submortoniceras*, *Eupachydiscus*, *Menuites*, *Baculites sulcatus*, *Damesites* and *Argentoscaphites*, indicating an earliest Campanian age.

Fossiliferous pebbles and cobbles of sandy limestone, similar to those at Mzamba, occur among the beach pebbles at the Mtata estuary to the south, pointing to offshore Cretaceous deposits at this locality.

Igneous activity

A cluster of kimberlites in East Griqualand include the **Melkfontein Carbonatite Tuff** with an age of 63.4Ma (earliest Palaeocene). They suggest local igneous activity in the area at the Cretaceous/Tertiary boundary.

KwaZulu-Natal

Cretaceous rocks in southern KwaZulu-Natal are restricted to a few small coastal exposures of Early Campanian age.

Mzamba Formation

Small outliers of the Mzamba Formation occur between the Inkandandhlovu and Itongazi rivers, and again near the Mpenjati estuary. At the latter a basal conglomerate rests on an uneven surface of Margate granite, and is overlain by shellbeds alternating with calcareous sandstones. Strata exposed in the surf zone at low tide are rich in *Teredo*-bored logs, and abundant vertebrate remains which include the plesiosaur *Cimoliasaurus*, the mososaur *Liodon*, sharks' teeth assigned to *Squalicorax*, *Cestracion*, *Cretoxyrhina*, *Cretolamna*, *Odontaspis* and *Scapanorhynchus*, and turtle fragments. Also present is the bivalve *Linotrigonia plumosa* which occurs in the earliest Campanian at the Wild Coast casino.

Umbilo Formation

Excavations at Durban have proven the presence of subsurface Cretaceous strata resting unconformably on Dwyka (Permo-Carboniferous) glacigenes. The discovery of the ammonite *Gunnarites kalika* in remanié at the base of the unconformably-overlying Bluff Formation suggests an almost complete offshore succession up to the Late Maastrichtian.

The base of the Umbilo is marked by a thin polymictic conglomerate with well-worn cobbles of Natal sandstone, Dwyka diamictite, Ecca lydianite, Karoo dolerite and granite-gneiss. The overlying succession attains a thickness of >100 m and comprises pyritic black, grey and yellow shales and glauconitic siltstones with interbeds of fine-grained sandstone, shelly limestone and concretionary horizons. There is extensive phosphatization and local pyritization of faunas.

Excavations at Sometsu Road and Maydon Wharf yielded the ammonites *Texanites*, *Submortoniceras*, *Hoplitoplacenticeras*, *Eupachydiscus*, *Natalites*, *Baculites*, *Gardeniceras* and *Anagaudryceras*, together with the bivalves *Inoceramus* and *Phelopteria* and the limpet *Diodora*. The fauna is mostly Early to Middle Campanian.

Zululand

The most extensive exposure of Cretaceous rocks in Southern Africa occurs on the

Maputaland coastal plain. Initially deposition was mostly volcanic and fluviatile, but with rising 1st-order sea-levels the coastal plain was inundated and magmatism ceased.

Bumbeni Complex

The Bumbeni is a terrestrial A-type igneous complex at the southern end of the Ubombo Range, southeast of Jozini. Here it straddles the Msunduze R. and rests unconformably on mid-Jurassic rhyolites. At the base is the fluviatile **Msunduze Formation**, a unit of conglomerates, sandstones and arkoses of debris-flow origin. Waterworn volcanic clasts up to a metre in diameter are generally unsorted and mostly of local derivation. Fossil logs are common.

The overlying **Mpilo Formation** is exposed on the south flank of Mpilo Hill, represented by 50 m of amygdaloidal trachybasalts and trachyandesites. Correlation with the Movene basalts of southern Moçambique is suggested.

The **Nxwala Rhyolite** follows conformably on the Mpilo and comprises rhyolitic pitchstones and volcaniclastics. At the base is a thin devitrified obsidian (pitchstone) with contorted flow banding which is exposed irregularly throughout the Nxwala area. Outcrops have the appearance of small and discontinuous domes, flows and lobes, pointing to low-viscosity eruptions.

Overlying the pitchstones are pyroclastics which include chaotic breccias, classic airfall tuffs and stacked beds of tephra. These document **plinian eruptions** such as the one witnessed by the Greek philosopher Pliny which devastated Pompeii in AD79. Eruptions record pressure build up beneath a blocked volcanic vent and explosive clearance of the conduit plug. The resulting fine pyroclastic debris is blasted kilometres into the air by a jet of gas before expanding into a cloud of magma globules and ash which then rains back to earth.

Following conformably on Nxwala pyroclastics are finely-banded and contorted porphyritic rhyolites of the **Fenda Formation**. South of the Nxwala perlite mine these form low domes intruding poorly-exposed tuffs, perlitic pitchstones and rhyolites. Some may be subtrusive bodies which failed to reach the surface. They are intruded by the plug-like **Ring Syenite** showing features of an alkaline ring complex.

Along the Munywane stream, and again NW of the Nxwala perlite mine, highly-altered green airfall tuffs of the **Golweni Formation** mark the base of the Bumbeni Complex. These are overlain by finely-banded rhyolites of the **Munywane Formation** with zones of auto-brecciation, flow folding, contorted flow banding and micro-faulting.

The **Kuleni Rhyolite**, of earliest Berriasian age (145 Ma), builds dome-shaped hills with distinctive vegetation. It occurs as irregular bodies intruding Jozini rhyolite and comprises flow-folded rhyolites which, like the Fenda, may be high-level subtrusives.

Zululand Supergroup

The most complete and extensive exposures of marine Cretaceous rocks in Southern Africa are found on the Maputaland coastal plain. Here the **Zululand Supergroup** follows unconformably on volcanics of the Lebombo Group and Bumbeni Complex, and is divided into:

ii.) **Maputaland Group** - a regressive succession of shallow-marine and backshore aeolian deposits of Eocene to Recent age.

i.) **Hluhluwe Group** - a transgressive phase of shallow- to deep-water, often glauconitic, concretionary siltstones and very fine-grained sandstones yielding Barremian to Palaeocene faunas.

Hluhluwe Group

Sediments of the **Hluhluwe Group** attain a thickness of about 1000 m in the vicinity of Lake St Lucia, from where they thicken eastwards and perhaps northwards as well. The succession is subdivided on the basis of internal regional discontinuities into a number of tectonostratigraphic packages:

vi.) **Richards Bay Formation** - deep-water marine siltstones with a Palaeocene microfauna.

v.) **St Lucia Formation** - concretionary siltstones and very fine-grained sandstones with Coniacian to Late Maastrichtian marine faunas. There may be minor internal discontinuities.

iv.) **Skoenberg Formation** - concretionary siltstones and shellbeds yielding Early to early Late Cenomanian marine faunas.

iii.) **Mzinene Formation** - concretionary siltstones and shellbeds yielding late Early Albian to Late Albian marine faunas.

ii.) **Ndabana Formation** - concretionary siltstones and shellbeds with Late Aptian marine fossils.

i.) **Makhatini Formation** - terrestrial polymictic conglomerates and fluviatile to estuarine sandstones which grade laterally and vertically into glauconitic fine sandstones with courses of concretions yielding Barremian and Early Aptian faunas.

Makhatini Formation

Coarse terrigenous clastics at the base of the formation comprise mainly poorly-exposed polymictic fluviatile conglomerates and sandstones with fossil logs. This terrestrial facies fines eastward (coastward) and vertically into estuarine and shallowmarine sandstones and siltstones with thin conglomerates, trigonia shellbeds and *Teredo*-bored logs.

Along the Zibayeni (= Mlambongwenya) stream, 28 km NNE of Otobotini, these passage beds are represented by bioturbated and weakly cross-bedded, greyish-buff siltstones with courses of concretions crowded with drifted molluscs and fossil logs. Some of the shell beds are so crammed with ammonites, mainly *Colchidites* and *Paraimerites*, as to form lumachelles, and large *Megacucullaea* in life position are conspicuous. The above ammonites, together with *Heteroceras*, *Hemihoplites*, *Sanmartinoceras*, *Eulytoceras* and *Cryptocrioceras* date the commencement of marine sedimentation to the Barremian. The large thick-shelled oyster *Aetostreon latissimum* is particularly abundant, together with *Utrobiqueostreon*, *Pseudoyaadia hennigi*, *Mayesella*, *Megatrigonia*, *Austromyophorella* and the belemnite *Chalalabelus*. Abundant plant debris, logs and poor sorting of the sediments point to nearshore, low-energy, back-barrier sedimentation.

Early Aptian strata follow conformably on the Barremian and also are exposed along the Zibayeni. Here concretions littering the topsoil have yielded *Procheloniceras*, *Cheloniceras* and the giant heteromorphs *Tropaeum*, *Audouliceras*, *Australiceras* and *Coopericeras*, together with the bivalves *Zulutrigonia* and *Pratulum*. This level is present also along the Mayezela (= Mayesa) stream where the same *Pseudoyaadia*, *Mayesella* and *Utrobiqueostreon* occur together with frequent belemnites, but no ammonites.

Ndabana Formation

The Ndabana Formation follows paraconformably on the Makhatini, the base marked by a course of hiatus concretions which, along the Zibayeni, have been washed into a thick conglomerate. That the non-sequence spans much of the Middle Aptian is suggested by the absence of such diagnostic elements as *Dufrenoyia*, *Gargasiceras* and *Colombiceras* from the fauna.

Late Aptian strata are widely exposed in the Mkuze Game Reserve along the eastern and southern sides of Nhlohlela Pan, and again along the Mfolozi NNE of Otobotini. The concretionary siltstones are rich in the ammonites *Acanthohoplites*, *Parahoplites*, *Diadochoceras*, *Helicancyloceras*, *Tonohamites*, *Toxoceratoides*, huge *Lytoceras*, *Euphylloceras* and the bivalves *Megatrigonia*, *Pisotrigonia*, *Rinetrigonia*, *Nototrigonia* and *Sphenotrigonia*. At the Manyola Drift these beds have yielded the giant ammonite *Tropeaum*.

The uppermost beds of the Ndabana Formation are exposed a few hundred metres west of the Zibayeni store, just below the Aptian/Albian discontinuity. Here concretionary siltstones contain the ammonites *Nolaniceras* and *Epicheloniceras* indicative of a *subnodosocostatum* Zone age, with an abundance of brachiopods pointing to shoaling and marine regression. This points to latest Aptian regression and strata of this age are absent or not exposed in Zululand due to transgressive onlap by younger deposits.

Mzinene Formation

As in many parts of the world, the earliest Albian records a lowstand of sea level, and strata of this age are absent in Zululand. The non-sequence is marked by a course of hiatus concretions, above which is a drifted shellbed rich in the bivalves *Megatrigonia* and *Iotrigonia*, with the ammonites *Douvilleiceras* and *Tegoceras* serving to date the base of the formation to the late Early Albian (*steinmanni* Zone).

The strongly-transgressive nature of the Mzinene Formation is evident along the Sibica (= Mzinene) and Zibayeni streams where it rests disconformably on Late Aptian strata of the Ndabana Formation. On the road to Sodwana Bay it oversteps the latter to rest directly on Munywana rhyolites. Here the base of the formation is a yellow oyster-rich calcarenite with *Rastellum allobrogensis*, the ammonites *Douvilleiceras* and *Eubrancoceras*, and the gastropod *Pseudomelania*.

The remainder of the Middle and Late Albian comprises some of the most richly fossiliferous strata of this age known any where. The early Middle Albian is well represented by a diverse ammonite fauna including *Lyelliceras*, *Mirapelia*, *Eubrancoceras*, *Alopecoceras*, *Carinophylloceras*, *Ammoceratites*, *Anagaudryceras*, *Umsinenoceras* and *Rossalites*. Also present are the bivalves *Sphenotrigonia*, *Globocardium*, *Pterotrigonia*, *Neithea*, *Gyrostrea*, *Gervillella* and *Epicyprina*, diverse brachiopods, the nautiloid *Cymatoceras* and the echinoid *Hemiaster*. The later Middle Albian is poorly exposed but along the Mzinene R. has yielded *Mojsisovicsia*, *Mirapelia*, *Androiavites*, *Manuaniceras*, *Pseudhelicoceras* and *Euphylloceras*, together with the bivalves *Pterotrigonia* and *Ptilotrigonia*.

Near the old sisal factory, north of Hluhluwe, the Late Albian comprises bioturbated siltstones with thin concretionary shellbeds crowded with drifted molluscs and occasional pebbles. Its transgressive character is evident on the Makhatini Flats where it rests directly on coarse clastics of the Ndabana Formation. These marine beds, exposed in a quarry alongside the Sodwana Bay road, are crammed with a typical *cristatum* Zone fauna including the index species *Dipoloceras cristatum*, abundant *Manuaniceras*, as well as *Bhimaites*, *Hyporbulites*, *Hemiturrilites*, *Neophlycticeras* and *Hamites*, the bivalves *Neithea*, *Ptilotrigonia*, *Rutitrigonia*, *Rine-*

trigonia, Idonearca, Gervillella, Glycymeris, Epicyprina, Globocardium, Laevicardium, Protocardia, Procardia, Amphidonte, Gyrostrea and *Actinoceramus concentricus*, the gastropods *Turritella, Avellana, Confusiscala, Gyrodes* and *Mesoglauconia*, the scaphopod *Dentalium* and the echinoids *Hemiaster, Pseudholaster* and *Pygurus*. A minor break in the succession, marked by "nests" of bored concretions, splits the *cristatum* zone with *Hysteroceras* appearing in the upper part.

The best exposures of the *pricei, goodhalli* and *inflatum* Zones occurred along the Sibica, in stream bed and river bank sections 1-2 km ENE of the old sisal factory. The occurrence was made world famous by E. C. N. van Hoepen. Here the *pricei* Zone is marked by the incoming of *Deiradoceras, Mimeloceras, Drepanoceras, Rusoceras* and *Venezoliceras*; *Hysteroceras* is particularly abundant. The overlying strata range up into the *inflatum* Zone and are rich in the ammonites *Mortoniceras, Pervinquieria, Cainoceras, Letheceras, Arestoceras, Styphloceras, Pagoceras, Ameleceras* and many other mortoniceratines. The spectacular giant desmoceratid *Achilleoceras* is from this level, as is the lobster *Pseudohomarus* and the brachiopods *Cyclothyris, Cyrtothyris, Praelongithyris* and *Dzirulina*.

The *rostrata* Zone is not exposed in Zululand but a classic *perinflatus* Zone fauna occurs in a roadside gravel pit on the north bank of the Msunduze River at Ndumo. The ammonites include *Stoliczkaia, Paraturrilites, Durnovarites*, giant *Hypengonoceras* up to a metre in diameter, *Anisoceras perarmatum* and *Desmoceras*, together with the bivalves *Amphidonte, Pterotrigonia, Pleurotrigonia, Neithea, Inoceramus, Goniomya, Pleuromya* and *Gervillia*; also present are the echinoids *Hemiaster* and *Pygurus* and the brachiopod *Cyclothyris*. There is a marked change in the composition of the bivalve fauna at this level and its affinities are with the Cenomanian rather than the Albian.

The Albian-Cenomanian boundary is not exposed in Zululand, but at Ndumo strata with *D. perinflatus* pass upwards into a poorly-fossiliferous pebbly nearshore facies indicative of shoaling and regression. There is certainly a non-sequence, as in southern Moçambique, since the lowest Cenomanian strata at this locality are rich in the ammonite

Bored concretions

Early diagenetic concretions formed below the sediment-water interface may be disinterred by high-energy erosive events. They then litter the seafloor, wholly or partially exposed, to form a **discontinuous hardground** which provides hard rock surfaces for boring by lithodomous bivalves (*Botula, Lithophaga*) and polychaete worms, and encrustation by oysters, serpulid worms (*Proserpula, Spiroserpula*), calcareous algae and hermatypic corals. Such bored concretions document minor breaks in the stratigraphical succession.

Bored concretions with a multiple history of disinterment, boring and encrustation, reburial and renewed concretion growth are termed **hiatus concretions**. They mark significant stratigraphical discontinuities.

Mantelliceras saxbii (Sharpe) indicating a level well above the base of the Cenomanian and pointing to an hiatus spanning the *dispar*, *briacensis* and *lozoi* Zones. This discontinuity justifies resurrection of the Skoenberg Beds (= Formation) as a discrete tectonostratigraphical package.

Skoenberg Formation

Early Cenomanian strata are of limited extent in Zululand, mostly obscured by transgressive onlap by the St Lucia Formation. At Ndumo it is represented by about 100 m of siltstones with courses of fossiliferous concretions exposed in erosion gullies between the trading store and police station. Below the police station the fauna, preserved mainly as ochreous internal moulds, is dominated by the ammonites *Mantelliceras saxbii*, *Hypoturrilites* and *Mariella*, together with abundant *Pterotrigonia setosa* and *Trigonarca*. These are associated with *Goniomya*, *Pleuromya*, *Modiolus*, *Isognomon*, *Gervillia*, *Neithea* and *Inoceramus*, the nautiloid *Cymatoceras*, the echinoid *Hemiaster* and gastropods.

A similar fauna occurs immediately west of the celebrated Skoenberg locality, a crescentic meander scarp rising 30 m above the floodplain at the junction of the Sibica and Munywana (= Manuan) Rivers in the Phinda game reserve. Here the Early Cenomanian is only 10 m thick and comprises soft, decalcified, yellowish-buff bioturbated siltstones with scattered concretions alternating with hard, blocky to flaggy siltstone crowded with drifted and *in situ* molluscs. Besides those mentioned already, other ammonites include *Sharpeiceras*, *Forbesiceras* and the heteromorphs *Neostlingoceras*, *Ostlingoceras* and *Sciponoceras*.

Middle and early Late Cenomanian strata in Zululand are restricted to the Skoenberg where calcitized fossils can be picked up on the grassy slopes. The western end of the hill has yielded a typical *costatus* Subzone faunule with abundant *Turrilites costatus*, *Acanthoceras*, *Newboldiceras*, *Sharpeiceras* and *Hypoturrilites*, together with the bivalves *Pleurotrigonia*, *Pterotrigonia*, *Trigonarca* and *Goniomya*. Hermatypic corals (*Caryophyllia*, *Calamophyllia*, *Rhabdophyllia?*) are common at this level, and the echinoid *Periaster* occurs also.

Higher levels in the Middle Cenomanian, further to the east, have yielded *Acanthoceras*, *Gentoniceras* and *Newboldiceras* with which are associated *Turrilites acutus*, *Zelandites*, *Gaudryceras*, *Borissiakoceras* and *Phyllopachyceras*. Finds of *Pseudocalycoceras* and *Calycoceras* in surface scree point also to the presence of early Late Cenomanian (*guerangeri* Zone) strata.

The Cenomanian/Turonian disconformity

One hundred metres due north of the farm Belvedere, in the bed of the Sibica, are exposed yellow Cenomanian siltstones whose highest course of concretions is bored by the bivalve *Lithophaga*. The overlying hard calcareous sandstone is a nearshore lumachelle with abundant *Acanthotrigonia*, bivalve debris, pebbles and occasional *Proplacenticeras*. This hardground marks a period of non-deposition and erosion spanning much of the Late Cenomanian and Turonian.

Riverview Formation

Just east of Lake Mpangani, 2.5 km SW of Mtubatuba, poorly-sorted, polymictic, clast- supported conglomerates of the Riverview Formation fill a palaeovalley in Sabie River basalt. Clasts up to 0.7 m in diameter are predominantly basalt with subordinate rhyolite, dolerite, quartzite, granite, banded iron-formation and vein quartz. Much of the matrix derives from *in situ* decomposition of clasts. Crude bedding and low-angle cross-bedding is discernible. Black opalized logs were common, and an argillaceous lens has yielded leaves of cycads (*Otozamites*, *Pterophyllum*, *Taeniopteris*, *Dictyozamites*) and ferns tentatively identified as *Cladophlebis* and *Sphenopteris*. Unlike the Makhatini conglomerate basalt clasts prevail pointing to a long period of erosion and stripping of the Jozini rhyolites prior to deposition.

St Lucia Formation

Along the Sibica the St Lucia Formation follows disconformably on the Late Cenomanian hardground with a basal unit of buff, cross-bedded, fine-grained sandstones with occasional shell lentils, silicified logs and *Acanthotrigonia shepstonei*. Further south the formation transgressively oversteps older formations to rest directly on Precambrian basement at Richards Bay and Mtubatuba.

The age of the base of the formation has been the subject of some debate. Van Hoepen suggested, on the basis of misidentification, that Turonian strata were present along the Mkuze. Kennedy & Klinger dated the base as Early Coniacian on the basis of *Proplacenticeras*, *Forresteria* and *Kossmaticeras*. However all these genera are known now to extend into the Late Turonian, and the *Kossmaticeras* at this level is most like Late Turonian *K. recurrens*; more work is needed. Also present at this level are *Mesopuzosia*, *Eubostrychoceras* and the bivalve *Modiolus typicus*.

The base of the St Lucia Formation is again exposed on the eastern flank of Mkweyane (= Umkwelane) Hill, 1.1 km north of Haig Halt. Here it oversteps older Cretaceous strata to rest directly on a deeply-weathered, in places channelled, surface of Jurassic basalts, or even Archaean granite-gneiss of the Mpangeni Formation, the latter two formations juxtaposed by Late Jurassic displacement along the **Eteza Fault** totalling almost 5 km.

At a railway cutting just south of Haig Halt the base of the formation is a hard buff sandy limestone rich in the oyster *Ceratostreon* and spines of the echinoid *Dorocidaris*, with scattered pebbles of vein quartz and quartzite. The overlying concretionary siltstones, with lenses of shelly limestone representing riptide channel deposits, have yielded the ammonites *Peroniceras*, *Forresteria* and *Proplacenticeras* of Early to Middle Coniacian age.

Along the lower reaches of the Hluhluwe, on Glenpark Estates, the St Lucia Formation oversteps the Cenomanian to rest directly upon Late Albian strata and, on the north bank of the White Mfolosi, follows paraconformably on Riverview conglomerates.

The remainder of the St Lucia Formation is a monotonous succession of bioturbated glauconitic siltstones and very fine-grained sandstones with courses of concretions through to the Late Maastrichtian. Small-scale sedimentary cycles are conspicuous with erosively-based shellbeds, generally rich in pelletal glauconite in their lower part, fining upwards into massive siltstones with *in situ* faunas. The shellbeds, dominated by thick-shelled bivalves (oysters, trigonias), may record high-energy storm episodes (tempestites) with the siltstones representing background sedimentation in a low-energy shelf environment. Primary structures in the siltstones have been largely destroyed by intense bioturbation, but burrows of *Thalassinoides* and *Chondrites* are not uncommon.

The *petrocoriensis* and *tridorsatum* Zones are well represented along the north bank of the Sibica, immediately east of the road bridge. Here *Forresteria*, *Peroniceras* and *Proplacenticeras* abound, associated with *Kossmaticeras*, *Tongoboryceras*, *Allocrioceras*, *Baculites umsinenensis* and *Scaphites*. Also present are the bivalves *Modiolus*, *Veniella forbesiana*, *Nordenskjoeldia*, *Acanthotrigonia*, *Linotrigonia*, *Macrocallista*, *Labrostrea*, various inoceramids (*Inoceramus*, *Mytiloides*, *Cremnoceras*, *Volviceramus*), gastropods (*Actaeonina*, *Cylichna*, *Gyrodes*, *Mayeria*, *Perissoptera*) and sharks' teeth (*Cretolamna*, *Scapanorhynchus*).

The Late Coniacian is exposed along the lower reaches of the Hluhluwe and Nyalazi Rivers, yielding *Paratexanites*, *Protexanites*, *Zuluiceras* and *Cremnoceramus*. Gauthiericeratids abound in Late Coniacian strata at Mason's Camp, 3.5 km north of Picnic Point, on the western shore of False Bay, together with many small bivalves and gastropods in aragonite preservation.

The Coniacian-Santonian boundary is currently in a state of flux, drawn at substantially different levels dependent on the usage of ammonites or inoceramid bivalves. To date most workers have drawn this boundary at the uncoming of the ammonite *Texanites* and, for the present, this criterion is followed here since inoceramids are facies fossils, generally absent from shallow nearshore environments.

Early Santonian *vanhoepeni* Zone strata are variably exposed along much of the western shore of False Bay, but are particularly well displayed in foreshore sections 3.2 km north of Picnic Point. Here the Coniacian-Santonian boundary can be drawn, as a matter of convenience, at a flood of *Baculites capensis*, immediately above which occurs *Texanites vanhoepeni*. Slightly higher beds have yielded *Plesiotexanites*, *Pleurotexanites*, *Pseudoschloenbachia*, *Damesites* and diverse gastropods, bivalves and echinoids. Santonian strata are also well exposed along Die Rooiwalle, a cliff on the NE shore of False Bay, north of Lister's Point.

Evidence for Santonian transgression occurs at Richards Bay where the St Lucia Formation oversteps older units to rest nonconformably on unweathered Archaean gneisses. Boreholes have intersected at least 100 m of concretionary dark grey sandy siltstones with rare mudstone lenses and sandy limestones. Ammonites from near the base

Hardgrounds

Hardgrounds develop where unconsolidated sediment on the seafloor is swept away by current action, exposing a hard lithified substrate for encrustation and boring by marine organisms (oysters, pholadid bivalves, serpulid worms) and for submarine mineralization (phosphatization, glauconitization, dolomitization).

Modern hardgrounds develop as a thinly-cemented crust on unconsolidated sediment and stratigraphical studies indicate hardground formation may take only a few thousand years. There is a close genetic relationship between hardgrounds and condensed sequences. Whereas condensed sequences are the product of minimal sedimentation, hardgrounds result from nil or negative sedimentation (erosion). The two are often associated.

include *Baculites*, *Pseudoschloenbachia* and *Madagascarites*, with the overlying succession ranging up into the Late Maastrichtian with abundant *Pachydiscus*, *Eubaculites* and *Hoploscaphites*, together with *Desmophyllites* and *Menuites*.

Late Santonian/Early Campanian strata are exposed to the east of Mkweyane Hill, in a roadside quarry which has yielded *Gardeniceras*, *Submortoniceras*, *Madagascarites* and the bivalves *Platyceramus*, *Linotrigonia*, *Pleuromya* and *Protocardia*.

The Santonian-Campanian boundary is still ill-defined macrofaunally in south-central Africa. Cliff sections and foreshore exposures along the SE shore of False Bay, extending from 3.5-6 km north of the mouth of the Nyalazi R., seem to span this interval. Here *Texanites* and *Reginaites* are common, with frequent *Submortoniceras* in the upper beds.

The SW tip of the Nibela Peninsula has, over the years, been visited by many geologists and extensively collected. The fauna at the base is a typical *delawarensis* Zone assemblage with *Australiella*, *Delawarella*, large *Baculites vanhoepeni* and *Menuites*. Amongst the bivalves oysters are conspicuous (*Zulostrea*, *Amphidonte*, *Exogyra*), together with *Endocostea* and rarer *Linotrigonia*, *Nordenskjoeldia* and *Spondylus* in aragonite preservation, thin-shelled pectinoids and the brachiopod *Cyrtothyris*. The overlying *marroti* Zone has yielded *Hoplitoplacenticeras*, *Eupachydiscus*, *Globocardium* and echinoids. Glauconite rich bioturbated horizons are conspicuous and record omission surfaces.

In Zululand the latest Campanian (*hyatti* Zone) is not recorded and believed to be absent on faunal grounds, although physical evidence for such a break is lacking.

Maastrichtian strata are well exposed in the cliff section along the shore of Lake St Lucia to the north of Charter's Creek. However the most informative sections are a series of gravel pits and river section in the canefields SW of Monzi. Here a diagnostic ammonite fauna occurs with *Pachydiscus*, abundant *Eubaculites*, *Saghalinites*, *Pseudokossmaticeras*, *Epiphylloceras* and *Gunnarites kalika* together with *Trochoceramus*, numerous oysters (*Phygraea*, *Agerostrea*, *Velostreon*, *Acutostrea*), fragile pectinoids and the echinoids *Bolbaster*, *"Hemiaster"* and *Micraster*. Overlying this is a level with *Menuites fresvillensis*, then follows strata with *Pachydiscus* gr. *gollevillensis*, *Brahmaites* and *Diplomoceras*, capped by a stunted and pyritic (now limonitized) fauna with abundant *Hoploscaphites*. Youngest Cretaceous strata, exposed on Fanie's Island at the south end of Lake St Lucia, are dominated by inoceramid debris only.

Several breaks in the Maastrichtian succession of Maputaland have been claimed on microfaunal grounds, but these have not been demonstrated in the field. One such non-sequence is said to occupy the upper part of the *Globotruncana gansseri* foraminiferal zone and another is allegedly marked by the absence of the lowermost Maastrichtian *Globotruncana stuarti* Zone.

Bathymetry of the St Lucia Formation

Although the monotonous lithology of the Late Cretaceous succession in the False Bay region gives few indications of fluctuations in sea-level, these are recorded by depth related ostracod assemblages documented by R. V. Dingle. Thus shallow-water environments are characterized by a high (60-90%) cytheroidean content, their number declining in deeper environments.

A preponderance of smooth-shelled and blind ostracods (*Bythocypris*, *Bairdoppilata*, *Cytherella*, *Krithe*) are considered typical of fairly deep-water environments, such as encountered on the mid- to outer-shelf and upper continental slope. This association is accompanied by subordination of the total cytheroidean component, and its first appearance is known as the "**Bythocypris Line**". During St Lucia sedimentation the "Bythocypris Line" was exceeded during the late Early Campanian, and much of the Maastrichtian (= upper part of Campanian IV and all of Campanian V to Maastrichtian II of Kennedy & Klinger, 1977).

Deepest-water conditions (>500 m) in the basin, indicative of outer continental shelf/ upper slope environments, were attendant during deposition of the lower parts of Maastrichtian I and Maastrichtian II. This is the preferred environment of an ostracod assemblage characterized by *Krithe nibelaensis*, *Oertliella maastrichtia*, *Xestoleberis luciaensis* and *Bythocypris richardsbayensis*.

Moçambique

Cretaceous rocks are well represented in southern Moçambique, with Neocomian volcanics in the Zambesi valley and fossiliferous shallow-marine deposits underlying much of the coastal plain. Unfortunately the latter are largely obscured by younger cover.

Sena Subgroup

Non-marine Cretaceous exposures are splendidly developed in the cliffs of the Lupata Gorge where they constitute the Sena Subgroup of the **Lupata Group**, and may be subdivided as follows:

iii.) **Tambara Formation** - about 250 m of yellowish to red and brown unsorted conglomeratic sandstones with angular to subangular clasts in a calcareous matrix. Volcaniclastic interbeds are frequent.

ii.) **Sarula Formation** - about 300 m of alkaline lavas, mostly phonolites but also with rarer types and frequent volcaniclastic interbeds.

i.) **Sorodzi Formation** - up to 100 m of greyish unsorted pebbly sandstones and matrix- supported conglomerates with angular to subangular clasts in a matrix of calcareous sandstone. Pyroclastic debris increases in abundance upwards.

The **Sorodzi Formation** rests with marked erosional break on Late Jurassic columnar rhyolites. The sandstones vary from massive to flaggy and cross-bedded, with both calcareous and feldspathic lithologies represented. In the rudites, boulders of gneiss and Karoo lithologies up to a metre in diameter are not uncommon, although pebbles are mostly vein quartz. A small outlier of sandy red marl 5 km WSW of Chirunda Hill yielded the limb bone of a theropod dinosaur.

The Sorodzi grades into overlying **Sarula lavas** by a gradual increase in pyroclastics which, near the top, include volcanic bombs and huge masses of scoria. The pyroclastics are variably developed, with at least 70 m of red tuffs at Chiganga.

Sarula volcanics are best exposed in the Lupata Gorge where they comprise mostly dull greenish-black to olive, grey, brown and chocolate phonolites, together with rarer nepheline basalt, leucitophyre, kenyaite, tinguaite and blairmorite, as well as occasional flows of rhyolite and trachyte. The red to purple and pink pyroclastics include sandy tuffs, crystal tuffs and fine-grained lithic tuffs with fragments of rhyolite and pitchstone up to 2 cm in diameter. An old whole-rock determination of 115 Ma suggests an Aptian age.

Sandstones of the **Tambara Formation** contain pebbles of Sarula lava and a stratigraphical discontinuity separates the two. They contain sporadic interbeds of tuff and thin lava.

East of Mt Gorongoza alkaline lavas of the **Inhandue Formation** build the range of that name, as well as being responsible for Mts Bunga, Chiuata and Fuvo to the south. They comprise mainly trachytes and phonolites. A Late Hauterivian age (131 Ma) suggests contemporaneity with the Bumbeni Complex of Maputaland.

Away from Lupata gorge, Sarula lavas are no longer developed and the tripartite subdivision of the Sena breaks down. The unit then becomes the **Sena Sandstone Formation** and comprises mostly fluviatile sandstones, grits and conglomerates. It crops out at the base of the Cheringoma plateau, 60 km to the south, were it is overlain by strongly-transgressive Maastrichtian marine deposits of the Grudja Formation.

Maputo Formation

South of 23°S of latitude, the Sena Sandstone grades into marine deposits of the Maputo Formation of Barremian to Cenomanian age. It rests with slight angular discordance and a prominent basal conglomerate on deeply-weathered Lebombo rhyolites. Clasts in the basal conglomerate are mostly rhyolite, with some basalt and basement granite-gneiss.

Along the Maputo R., the basal conglomerate is overlain by 4-5 m of fine-grained sandstone with *Megatrigonia*. On the Xessane R., nodules in fine-grained glauconitic sandstones 3-4 m above the base have yielded a sparse Barremian faunule including the ammonite *Crioceratites*. Also from this level or the basal Aptian, recovered from a coal-prospecting shaft, are the bivalves *Trigonia*, *Coelastarte*, *Camptonectes*, *Crassatella*, *Entolium*, *Goniomya*, *Neithea*, *Panopea*, *Globocardium* and *Pseudopteria*, the gastropods *Anchura*, *Avellana*, *Gyrodes*, *"Pleurotomaria"* and *Ringinella* .

The Aptian of southern Moçambique forms a northward-thinning wedge from at least 200 m thick along the Maputo R. to only 120 m along the Xessane. There is no evidence for earliest Aptian strata, probably due to onlap by the later Early Aptian. The latter was a period of widespread transgression in southern Moçambique, in places lapping onto Lebombo volcanics. Strata of this age are the source of the

Omission surfaces

Burrowed horizons mark short breaks in sedimentation known as **omission surfaces.** Usually these are rendered conspicuous by different sediment types above and below the burrowed surface. Three categories of trace fossils are associated with omission surfaces:

The **pre-omission suite** of burrows records the activities of animals churning sediment prior to the hiatus. In general bioturbation is muted due to a lack of contrast between burrow matrix and infill.

The **omission suite** of burrows is superimposed on the pre-omission suite and is rendered conspicuous by contrasted fillings, cross-cutting relationships and by an increased density of burrows.

The **post-omission suite** of trace fossils records bioturbation following the recommencement of sedimentation, and is thus superimposed on the earlier two suites.

classic fossil assemblages from Chalala, Lubemba and Powell's Camp, dominated by *Cheloniceras* together with *Audouliceras*, *Australiceras*, *Lithancylus*, *Ammonitoceras*, *Aconeceras*, *Macroscaphites*, *Valdedorsella*, *Toxoceratoides*, *Sanmartinoceras* and *Euphylloceras*. At Chalala this part of the succession is represented by about 6 m of pale greyish-brown siltstones and sandstones in two fining-upward cycles, the base of each marked by a bed of grey sandy limestone.

The Late Aptian comprises fine-grained glauconitic and argillaceous sandstones, sandy limestones with courses of concretions yielding the ammonites *Acanthohoplites*, *Nolaniceras*, *Epicheloniceras*, *Tropaeum* and *Ammonitoceras*. Some of the calcarenites are capped by oyster-encrusted hardgrounds and mark minor discontinuities. One such hardground along the Maputo R., overlain by strata with *Acanthohoplites*, may mark the mid-Aptian non- sequence identified in Maputaland. Pebbly sandstones, conglomerates and thick-shelled bivalves indicative of nearshore sedimentation document terminal Aptian regression.

The Aptian/Albian boundary is a non-sequence in southern Moçambique, as in Maputaland. Northeast of Chalala, the Late Albian transgressively oversteps older strata to rest discordantly on the Early Aptian with a basal conglomerate. A float specimen of *Mirapelia* is the only evidence for the presence of Middle Albian strata.

The Late Albian is well represented at Catuane where strata of the *cristatum* and *pricei* Zones are exposed and, unsurprisingly, are similar to those from Maputaland. They have yielded *Dipoloceras*, *Drepanoceras*, *Deiradoceras*, *Mortoniceras*, *Hysteroceras*, *Myloceras*, *Labeceras*, *Venezoliceras*, *Beudanticeras*, *Anisoceras* and *Tetragonites*.

Pebble beds and conglomerates characterize the latest Albian along the Maputo River and, as in Maputaland, provide evidence for terminal Albian regression and a stratigraphical discontinuity.

Basal Cenomanian strata are absent at outcrop in southern Moçambique, the non-sequence marked by a level of hiatus concretions overlain by comminuted echinoid debris passing upwards into an *in situ* echinoid fauna. Overlying strata exposed along the Coane R. have yielded an Early Cenomanian fauna including the ammonites *Mantelliceras saxbii*, *Sharpeiceras*, *Mariella*, *Ostlingoceras* and *Sciponoceras*, together with the bivalves *Pterotrigonia*, *Pleuromya* and *Goniomya*, the gastropods *Avellana*, *Latiala* and *Confusiscala*, rhynchonellid brachiopods and the echinoids *Hemiaster* and *Pseudholaster*.

The earliest Middle Cenomanian (*costatus* Subzone) is strongly transgressive along the Mahube R. where it rests on Early Aptian strata with a *Lithophaga*-bored hardground. The fauna is like that in Maputaland and includes *Acanthoceras*, *Newboldiceras*, *Turrilites costatus*, *Cenomariella*, *Sciponoceras*, *Puzosia* and *Tetragonites*, the bivalves *Pleurotrigonia*, *Rutitrigonia*, *Trigonarca* and *Aucellina*, diverse gastropods, the echinoid *Douvillaster* and the crab *Paranecrocarcinus*.

Boane Formation

There is no evidence for outcrop exposures of Late Cenomanian, Turonian or Coniacian marine deposits in southern Moçambique. Whereas the absence of the Late Cenomanian and Turonian may be, as in Maputaland, due to a non-sequence, it is likely that surface exposures of the Coniacian are obscured by transgressive onlap of the overlying Grudja Formation.

Unfossiliferous fluviatile clastics of the Boane are thought to be Turonian, thus corresponding with the Riverview Formation of Maputaland. Along the Umbeluzi R., to the SW of Maputo, this unit is at least 150 m thick and comprises poorly-sorted, cream, medium- to coarse-grained arkoses, red fine-grained arkoses with convolute lamination, laminated claystones, khaki sandstones with pebble lags, and conglomerates consisting almost entirely of well-rounded rhyolite pebbles.

Domo Formation

The Domo Formation is known only from the subsurface, between the Zambesi and 25°S of latitude, where it follows on either the Maputo or Sena Formations with a gradational contact. It was deposited in an elongate NS-trending trough, an extension of the **Chissenga Graben**, and grades westwards into Sena sandstone.

The Domo comprises 325-1522 m of dark grey to black, thin-bedded marly shales with interbeds of sandstone. These lithologies, together with a stunted molluscan fauna in the lower part, suggest euxinic sedimentation in a barred trough, with planktonic foraminifers documenting periodic influxes of marine water. The Domo Shale is dated microfaunally as Albian to Turonian.

Grudja Formation

As along the KwaZulu-Natal coast, the Early Campanian is strongly transgressive in southern Moçambique, initiating deposition of the Grudja Formation.

At Madebula, 700 m SW of the Maçandene trig beacon, this formation comprises yellowish fine-grained glauconitic sandstones with subrounded chalcedony grains and pebbles washes of well-rounded rhyolite. The rich invertebrate fauna is similar to that at Mzamba and includes the Early Campanian ammonite *Baculites boulei*, together with thc bivalves *Acanthotrigonia*, *Camptonectes*, *Crassatella*, *Eriphyla*, *Glycymeris*, *Platyceramus*, *Linotrigonia*, *Mesocallista*, *Neithea*, *Nordenskjoeldia*, *Paraesa*, *Pleuromya*, *Protocardia* and *Veniella*, and the gastropods *Avellana*, *Cryptorhytis*, *Gyrodes*, *Trochactaeon*, *Pugnellus* and *Volutilithes*. Grudja strata are exposed also 3 km south of the bridge across the Porto Henrique-Bela Vista road, and at Umbeluzi.

Late Campanian strata are not recorded from southern Moçambique, probably due to the strongly transgressive nature of the overlying Maastrichtian. The latter represents the period of peak transgression during the Cretaceous of this region.

Cretaceous exposures between the Zambesi and Save Rivers are confined mainly to outcrops in the Buzi valley and the western face of the 300 m high Cheringome plateau. In the Mazamba R., which incises the western face of the Cheringome, the Cretaceous is capped by Eocene nummulitic limestones and comprises:

iii.) 65 m of cream fossiliferous limestones with *Eubaculites* in the upper part.

ii.) 16 m of fossiliferous grey sandy clays, reddish and mottled near the top.

i.) 16 m of unfossiliferous red pebbly grit.

The clays of unit (ii) are also exposed in the Kundwi R. where again they have yielded *Eubaculites* together with the nautiloids *Hercoglossa* and *Eutrephoceras*. Then follows 67 m of grey to yellow sandy clays which pass upwards into 80 m of sandy glauconitic limestones.

In the Buzi valley, the Grudja attains a thickness of 91 m with prime exposures in the cliff sections at Maxemeje. Here at least 30 m of highly fossiliferous, compact, cream limestones are exposed, in places sandy and friable. Near the middle is an oyster band rich in *Agerostrea ungulata*.

In the Inchaninga area, the Grudja is only 18 m thick but boreholes show it to thicken markedly eastwards to over 1520 m in coastal wells where a continuous Coniacian to Palaeocene sequence is claimed.

Incomanine Formation

At Incomanine, on the east bank of Lake Chibamassala, 10 km NE of Sabie, yellowish conglomerates, pebbly sandstones and coarse-grained calcareous sandstones of the Incomanine Formation (= Lourenço Marques Conglomerate) represent a shoreline facies of the terrestrial Malvernia Formation. They have yielded a rich invertebrate fauna including the bivalves *Incomatiella*, *Trigonocallista*, *Linearia*, *Venericardia*, *Glycymeris*, *Camptonectes*, *Lycettia*, *Nucula*, *Nuculana* and "*Cardium*", together with the gastropods *Euspira*, *Tudicla*, *Haustator*, *Confusiscala*, *Cerithidea*, *Pollia*?, *Littorinopsis*, *Tropidothais* and *Theodoxus*?.

Northern Moçambique

The Cretaceous rocks of northeastern Moçambique are not well known. Late Valanginian-Hauterivian strata are exposed on the Fernão Veloso peninsula, and the vicinity of Moçambique and Mahiba Hill, where they have yielded the ammonites *Olcostephanus*, *Lytoceras*, *Euphylloceras*, *Bochianites* and an unidentified neocomitid ammonite, together with the bivalves *Pleuromya* and *Thetis*.

Ammonites "resembling Aptian and Albian forms" are present at Mahiba Hill, west of Porto Ameli, and at Conducia the Late Albian has yielded the ammonites *Goodhallites*, *Mariella*, *Desmoceras*, *Beudanticeras*, *Anagaudryceras*, *Euphylloceras* and *Lytodiscoides*. The presence of the Early and Middle Cenomanian is indicated by the occurrence of *Sharpeiceras* and *Newboldiceras*. Senonian fossils, including the oyster *Agerostrea*, are reported from the environs of Conducia.

Zimbabwe

Cretaceous rocks in Zimbabwe are wholly terrestrial and mostly volcanic, both extrusive and intrusive (mainly kimberlites).

Volcanics in the Zambesi Valley

The **Chidziwa Formation** crops out in the foothills of the Zambesi Escarpment, south of Mkumbura, but is best developed in the Janika Hills north of the Zambesi. The succession comprises mainly dense, aphanitic, lilac trachytes which, with the appearance of plagioclase phenocrysts, grade into porphyries, together with amygdaloidal lavas, breccias and tuffs. The formation is correlated with the Early Cretaceous alkaline volcanics of the Lupata Group further to the southeast.

The **Chiswiti Formation** forms rugged topography in the Chiswiti Communal Area of NE Zimbabwe, to the west of the Senga R. in the Zambesi Graben. The volcanics, which have suffered extensive hydrothermal alteration, are downthrown along a major fault followed by the Senga, and appear to overlie Karoo (Triassic) sediments unconformably. Well-laminated fine-grained flows, some with vesicular flowtops and ropy surfaces, are present as well as agglomerates, ignimbrites and white potash-rich pumice. Eruption is believed to relate to Lupata volcanism.

Malvernia Formation

In the extreme SE of Zimbabwe, post-Karoo sediments of the **Malvernia Formation** build the Tjolotjo Cliffs along the banks of the Lundi R. in the Ghona-re-Zhou. Here the 300 m-thick succession is divided into a lower member of pale conglomerates with an extremely calcareous, white basal conglomerate, and an upper member of red sandstones. Pebbles, cobbles and even boulders of all important Karoo lithologies occur abundantly in the conglomerates. In the more-northerly exposures pebble analysis suggests a northerly provenance but, in the south, abundant pebbles of Soutpansberg quartzite testify to a westerly source.

Limited exposures of the Malvernia Formation again crop out along the Limpopo/Moçambique border, between Pafuri and Shingomeni. Here about 100 m of conglomerates and sandstones rest unconformably on Lebombo volcanics, the rudites including clasts of Soutpansberg and Clarens sandstone as well as Lebombo basalt up to 15 cm in diameter. These are overlain by resistant, cliff-forming, pink to white sandstones, in places calcareous and occasionally marly.

Although the Malvernia Formation has yielded only indeterminate bone and plant scraps, it appears to grade eastwards into the Maastrichtian Incomanine Formation of southern Moçambique.

Kalahari Group

A Cretaceous age has been suggested for silicified lacustrine deposit at the base of the Kalahari Group in western Zimbabwe but, at present, there is no way of dating these beds precisely. They have yielded only freshwater snails (*Limnaea, Bulinus*) and charophytes (*Chara*).

Angola

Marine Cretaceous rocks are well developed along the Angolan littoral but much of the exposure is obscured by thick bush and younger cover. There are two depositories, the Namibe (= Moçamedes) and Kwanza (= Cuanza) basins, the latter by far the larger.

Namibe basin

The terrestrial and shallow-marine sediments of the Namibe basin are assigned to the **Bentiaba** (= São Nicolau) **Group** within which the following subdivisions have been recognized:

vi.) **Mocuio Formation** - siltstones with a basal bone bed rich in Maastrichtian vertebrates.

v.) **Baba Formation** -Santonian marine limestones and nodular siltstones coarsening upwards into conglomerate.

iv.) **Ombe Volcanic Formation** - thin flows of basaltic and andesitic lava.

iii.) **Salinas Formation** - marine limestones with a Late Cenomanian-Early Coniacian fauna.

ii.) **Piambo Formation** - terrestrial conglomerates and red sandstones in a fining-upward sequence.

i.) **Cangulo Formation** - terrestrial siliciclastics, lacustrine limestones and evaporites.

The unfossiliferous **Cangulo Formation** is best exposed south of Chapéu Armado where it rests nonconformably on crystalline basement.At the base is the **Ponta Negra Member**, a heterolithic unit of sandstones, siltstones and shales with interbeds of gypsum and bituminous limestone. These are overlain conformably by the **Mocungo Member**, a thin siliceous dolomitic limestone impregnated with oxides of Mn, Ba, Fe and Ce, overlain by 30 m of manganiferous sandstones. A Neocomian-Aptian age is suggested.

The **Piambo Formation** follows disconformably on the Cangulo. At the base is the **Giraul Conglomerate** comprising up to

150 m of torrential fanconglomerates and fluviatile sandstones. These fine upwards into the **Ponta Grossa Member**, a unit of at least 40 m of red sandstones and variegated mudrocks. At the mouth of the Inamagando R. a marine intercalation of oolitic and oncolitic limestone, the **Inamagando Beds**, has yielded the Albian ammonite *Desmoceras*.

The **Salinas Formation** follows conformably on Ponta Grossa redbeds and records Late Cenomanian transgression. Grey nodular limestones at the base are crowded with the oyster *Rhynchostreon* and yield a classic *gracile* Zone fauna including the ammonites *Calycoceras*, *Kanabiceras*, *Pseudocalycoceras*, *Metoicoceras*, *Anapuzosia*, *Nigericeras* and *Watinoceras*, together with gastropods (*Pseudomelania*, *Turris*, *Gymnarus*, *Pterodonta?*) and bivalves (*Veniella*, *Trigonarca*, *Neithea*, *Psilomya*, *Protocardia*, *Costagyra*).

In the sea cliff 5 km due west of Bentiaba (= Posto do São Nicolau) a condensed sequence of alternating fossiliferous limestones and siltstones is exposed, with a bed of stromatolitic limestone at the base. Ammonites from the lower to middle part indicate an Early to Middle Turonian age and include the ammonites *Morrowites*, *Prionocyclus*, *Vascoceras*, *Mossamedites*, *Proplacenticeras*, *Puebloites*, *Baculites*, *Anagaudryceras* and *Damesites*. This same fossiliferous horizon is exposed below the Farol de Ponta Grossa. The upper part of the formation has yielded Early Coniacian ammonites including *Prionocycloceras*, *Forresteria*, *Kossmaticeras* and *Mesopuzosia*.

The **Ombe Formation** rests disconformably on baked and recrystallized Salinas limestones, and comprises thin flows of basalt, analcime basanite and andesite of Middle to Late Coniacian age. In the Jombovacai valley and the Damba do Ombe up to 3 phases of eruption are recorded by sedimentary interbeds.

The **Baba Formation** follows disconformably on Ombe lavas with a basal sedimentary breccia of volcanic clasts overlain by a massive grey limestone rich in the oyster *Phygraea*. Then follows a Middle Santonian succession of thin limestones and nodular siltstones with the ammonites *Texanites*, *?Protexanites*, *Gardeniceras*, *Menuites*, *Damesites* and *Baculites*, as well as the bivalves *Trachycardium*, *Veniella*, *Neithea*, *Protocardia*, *Granocardium*, *Oscillopha*, *Trigonarca*, *Crassatellites* and *Plicatula*, the echinoid *Bolbaster*, and the gastropod *Haustator*. Late Santonian-Early Campanian shoaling is indicated by the conformably overlying **Bero Conglomerate**, a proximal shoreline facies. Whether or not this formation extends elsewhere into the Campanian remains to be determined but, it seems, much of the Campanian is absent.

At Bentiaba the **Mocuio Formation** rests disconformably on the Baba with a 2 m-thick basal bone bed extremely rich in selachian teeth (*Squalicorax*, *Cretolamna*, *Carcharias*, *Rhombodus*), with which are associated the mosasaurs *Mosasaurus*, *Globidens*, *Prognathodon*, *Phosphorosaurus*, *Halisaurus* and "*Platycarpus*", the turtles *Toxochelys*, *Protostega* and *Euclastes?*, a plesiosaur close to *Tourangisaurus* and unidentified dinosaur and pterosaur remains. The Mocuio bone bed testifies to the existence of the **Beguella Current** and the high primary productivity associated with its upwelling. The overlying greenish siltstones and fine sandstones contain scattered vertebrate remains, limonitized burrow fills and occasional phosphatic concretions but an invertebrate fauna is lacking.

Kwanza basin

Understanding of the Cretaceous geology of the Kwanza basin is hampered by thick bush, poor outcrop, a veneer of Cainozoic strata, complex faulting resulting from post-Aptian salt tectonics, and westward tilting. The succession is assigned to the **Kwanza Group** but formations are poorly defined and limits imprecise due to poor access and a lack of outcrop continuity. Within the basin the southern part is separated as the Benguela Subbasin.

Benguela Subbasin

In this subbasin the base of the Cretaceous is formed by the **Dombe Formation**, a unit of coarse terrestrial clastics and lagoonal sediments with evaporites. Locally at the very base are beds of siliceous dolomitic limestone impregnated with Cu oxides and perhaps to be correlated with the Mocungo Member of the Namibe basin.Then follows oolitic limestones and calcarenites of the "**Formação com *Pholadomya***" with the bivalves *Pholadomya pleuromyaeformis*, *Panopea*, *Neithea* and *Lithophaga*, the gastropods *Retusa*, *Glauconia?*, *Actaeonella*, *Tylostoma*, *Pseudamaura* and the echinoids *Salenia* and *Pygurus*. A stratigraphical discontinuity separates the overlying brightly-coloured red sandstones and siltstones of the "**Formação com *Nerinea***" which were deposited on tidal flats. Besides *Eunerinea* the formation has yielded an indeterminate ammonite, and an Aptian age is inferred.

The **Cuio Formation** follows, perhaps disconformably, on the underlying redbeds and comprises limestones and calcareous siltstones with a late Early to early Middle Albian fauna including the ammonites *Douvilleiceras*, *Mirapelia*, *Manuaniceras*, *Tegoceras*, *Beudanticeras* and *Puzosia*, together with the echinoids *Holaster*, *Hemiaster* and *Leymeriaster*. Upwards this unit becomes increasingly arenaceous, with unfossiliferous grits and calcarenites spanning the later Middle and earliest Late Albian. Overlying bioclastic and oolitic calcarenites have yielded a Late Albian fauna, the lower part characterized by *Mortoniceras stoliczkai*, overlain by a level with *Hysteroceras choffati* and capped by strata with *Rusoceras*, *Prohysteroceras* and *Goodhallites* suggesting correlation with the *pricei/varicosum* Zones. Echinoids abound and include *Phyllobrissus*, *Acriaster*, *Plagiochasma*, *Temnocidaris*, *Cardiaster*, *Mecaster* and many others.

The **Coporolo Formation** follows disconformably on the Cuio and comprises brightly coloured conglomerates and gritty sandstones with the gastropods *Actaeonella*, *Cerithium*, *Trajanella*, *Avellana* and *Eunerinea*, the oyster *Ceratostreon*, and the echinoid *Stigmatopygus*. It is assigned to the Early Cenomanian but diagnostic ammonites are lacking.

Coporolo clastics are overlain disconformably by an undivided sedimentary succession. Thin limestones at the base are assigned to the Late Cenomanian, but Turonian and Coniacian strata are unrecorded and a bivalve faunule has a Santonian aspect.

The **Mocuio Formation** is present in this subbasin, yielding a Maastrichtian vertebrate assemblage similar to that of the Namibe basin but more diverse, the chondrichthyes including *Angolaia*, *Angolabatis*, *Clamydoselachus*, *Hexanchus*, *Pseudocorax* and many others.

Main Kwanza basin

In this basin much of the information on the Cretaceous comes from petroleum boreholes. Here the **Kwanza Group** rests unconformably on crystalline basement, with initial deposition in a number of independent grabens and half-grabens. Initial basin fill was of volcanics (**Flamingo Formation**) and a variety of fluviatile and lacustrine sediments including evaporites. These are dated palynologically to the Berriasian-Early Aptian.

Thermal subsidence and development of a sag basin initiated deposition of terrigenous siliciclastics. At the base are lacustrine sandstones of the **Maculungo Formation**, disconformably overlain by red argillaceous sandstones of the **Cuvo Formation** for which a Barremian age is suggested. Then follow dolomitic and calcareous sandstones of the **Chela Sandstone**, fining upwards into silty dolomites and marls. Some of the carbonates are rich in ostracods and freshwater bivalves, but also have yielded a marine bivalve tentatively identified as *Pseudaphrodina*. The fern *Pachypteris* has been obtained from thin coal seams.

The **Loeme Salt Formation** is a mid-Aptian evaporitic sequence with an average original thickness of 350 m. Later halokinesis has resulted in lateral migration and the build-up of the enormous salt diapirs of the Galinda antecline. In the Cabo Ledo area these evaporites are overlain by a subsurface carbonate reef facies, the **Quiango Formation**, but faunas are not reported.

Except in the Cabo Ledo area the **Binga Formation** follows conformably on Loeme evaporites. It is a Bahamas-type carbonate-shoal deposit of oolitic and bioclastic calcarenites and lithographic and bituminous limestones with the tintinnid *Calpionella* and the molluscs *Pinna*, *Panis* and *Nerita*. It is the main petroliferous unit in the basin.

The **Tuenza Formation** follows on Binga limestones and is a thick succession of lagoonal clastics and evaporites. The salts are stacked in evaporative cycles, halite and anhydrite at the base, followed by anhydrite and then dolomite. Eastwards, along the basin margin, this formation intertongues with terrigenous clastics of the **Dondo Formation** and westward, adjacent to the Longa uplift (a topographical high), grades into bioclastic and reefal limestones of the **Longa Formation** with evaporitic interbeds.

The **Cabo Ledo Formation** follows conformably on the Tuenza and is a marine carbonate sequence up to 850 m thick. At the base is the **Catumbela Member** comprising some 10-20 m of pisolitic bioclastic calcarenites and caprinoidal limestones with stromatolites and corals. The overlying **Quissonde Member** consists of rhythmic alternations of argillaceous limestone and shale becomings increasingly argillaceous upwards and with a rich and diverse Late Albian macrofauna. Endemicity of many of the ammonites has led to a somewhat localized biostratigraphy:

- *Stoliczkaia dispar* Zone
- *Durnovarites perinflatus* Zone
- *Subschloenbachia rostrata* Zone
- *Elobiceras conditum* Zone
- *Prohysteroceras wordiei* Zone
- *Neokentroceras curvicornu* Zone

Effectively but not precisely the *conditum* Zone correlates with the *inflata* Zone of Western Europe, the *wordiei* Zone with the *goodhalli* Zone, and the *curvicornu* Zone with the *pricei* Zone. The presence of the *cristatum* Zone in the Kwanza basin is based on very small juvenile dipoloceratids.

Cliff sections at the mouth of the Hanha R. are the source of the celebrated Albian fauna described by Otto Haas. Although a detailed biostratigraphy is not available the diverse fauna includes *Pervinquieria*, *Elobiceras*,

Lobitoceras, *Boeseites*, *Hysteroceras*, *Beudanticeras*, *Tetragonites*, *Puzosia*, *Eogaudryceras*, *Anisoceras*, *Idiohamites*, *Ptychoceras* and *Hamites*. Palaeontologically it exposes the *curvicornu* to *conditum* Zones.

The *perinflatus* and *dispar* Zones are present at Porto Amboim, yielding a rich ammonite fauna including *Stoliczkaia*, *Durnovarites*, *Drakeoceras*, *Reyreiceras*, *Paraturrilites*, *Tetragonites*, *Anisoceras*, *Idiohamites* and *Hyporbulites*. The uppermost Albian (?*briacensis* Zone) of the Sumbe area has yielded the endemic mortoniceratine *Moutaiceras*.

The **Sumbe** (= Novo Redondo) **Formation** follows on the Cabo Ledo Formation but the nature of the basal contact is unknown. Along the eastern margin of the basin the formation oversteps older units ultimately to rest on basement granite-gneiss. The occurrence of *Mantelliceras* cf. *saxbii* to the south of Porto Amboim indicates the presence of the Early Cenomanian. At Sumbe dark grey shales around the base of the chapel (?now demolished) yielded *Turrilites costatus* and pass upwards into alternating limestones and siltstones with a Middle Cenomanian fauna including *Cunningtoniceras*, *Texacanthoceras*, *Conlinoceras*, *Eucalycoceras*, *Turrilites acutus* and huge *Forbesiceras*, with the echinoids *Pygopistes* and *Coenholectypus*, and the oyster *Costagyra*. The occurrence of Late Cenomanian and Early Turonian in the same area is indicated by *Vascoceras*, *Kanabiceras* and *Pseudaspidoceras*.

The **Itombe Formation** (= Cacoba Formation) is a fining-upward sequence of brownish to yellowish sandstones, siltstones and mudstones which, at Cocaba, have yielded Late Turonian *Coilopoceras* and Early Coniacian *Hemitissotia*. These are associated with the renown **Iembe fauna** of vertebrates which includes the mosasaurs *Angolasaurus* and *Tylosaurus*, the dinosaur *Angolatitan*, the turtle *Angolachelys*, the selachians *Scapanorhynchus*, *Ptychodus*, *Hexanchus*, *Squalicorax*, *Cretoxyrhina*, *Cretodus*, *Cretoxyrhinus*, *Paranomotodon* and *Onchosaurus*, and the bony fish *Enchodus*. At Cabo Ledo a crushed Turonian fauna in dark grey sandstones and siltstones includes *Subprionocyclus*, *Romaniceras*, *Borrisiakoceras* and *Puebloites*, with overlying (?Coniacian) limestones yielding *Parapuzosia*.

Coniacian strata in the Kwanza basin are assigned to the **Pambala Formation**. At Massingano the presence of *Proplacenticeras* in association with the bivalves *Veniella forbesiana*, *Pholadomya*, *Acanthocardia*, *Venilicardia*, *Plicatula* and *Cyprimeria*?, and the gastropods *Nerita* and *Cryptorhytis*, are of this age. Elsewhere in the basin this formation has yielded a Middle Coniacian *Peroniceras dravidicum*.

The poorly-known **N'Golome Shale** is of Santonian to Early Campanian age. Along the Lifune R., north of Luanda, it has yielded a crushed Early Campanian faunule including *Texanites*, *Submortoniceras*, *Eulophoceras* and *Baculites boulei*.

Three kilometres to the north of the small fishing village of Praia-Egito, about 70 km north of Lobito, is an outlier of Middle Campanian strata, the **Quimbala Formation**, resting unconformably on the Albian. It is a coarsening-upward clastic sequence, with coarse-grained sandstones yielding *Hoplitoplacenticeras*, *Australiella*, *Kitchinites*, *Polyptychoceras*, *Didymoceras*, *Tetragonites*, *Manambolites* and *Epiphylloceras*, the echinoids *Leiostomaster* and *Tholaster*, and diverse gastropods.

Late Campanian strata are well exposed in the sea cliffs at Barra do Dande where dark shales of the **Teba Formation** with

Manambolites are conformably overlain by thinly-bedded alternations of yellowish marls, calcarenite, sandy limestone and inoceramite of the **Rio Dande Formation**. The latter has yielded a classic *hyatti* Zone fauna including *Nostoceras hyatti* and *N. helicinum*, together with *Phylloptychoceras*, *Neophylloceras*, *Manambolites*, *Didymoceras*, *Baculites* and the bivalve *"Inoceramus" langi*. That the succession extends well into the Maastrichtian is indicated by the presence also of the ammonite *Sphenodiscus*. A like fauna occurs inland, near Cabinda, and includes also *Menuites*, *Libycoceras*, *Solenoceras*, *Axonoceras* and *Baculites* gr. *anceps*.

Continental deposits

Terrestrial deposits of Cretaceous age cover wide area of northeastern Angola. At the base, resting unconformably on older rocks, is the **Continental Intercalaire**, a fluviolacustrine sequence of mudstones, sandstones and conglomerates of Aptian to Albian age. These are overlain disconformably by the **Calonda Formation** (part of the **Kwango Group**), which is dated to the Cenomanian by palynomorphs and fossil fish. It is a faultbound trough accumulation of interbedded sandstones, shales and conglomerates 40-60 m thick. Coarse fanglomerates at the top are diamondiferous.

Kimberlites

Hundreds of kimberlite pipes and diatremes occur in the Lunda and Moxico provinces of eastern Angola. Although not well dated most are considered Early Cretaceous, the **Catoca kimberlite** of NE Angola having yielded a zircon age of 117 Ma (mid-Aptian). Many of the pipes are covered by kimberlitic grits of the Calonda Formation.

MALAWI

Post-Karoo sediments of the **Karonga Formation** (= Dinosaur Beds) are exposed near Mwakasyunguti in the Karonga district of northern Malawi. These fill a broad depression in Karoo rocks, locally overstepping the latter to rest directly on crystalline basement, and comprise about 350 m of cross-bedded fluviatile sandstones and mudstones, together with lacustrine marls. The fauna includes the titanosaurid dinosaurs *Malawisaurus* and *Karongasaurus*, the crocodile *Malawisuchus*, the terrapin *Platycheloides* and the ostracod *Cypridea*. Although not precisely dated, they are thought to be Middle or Late Cretaceous.

ZAMBIA

Cretaceous rocks in Zambia are represented only by diamondiferous kimberlites of the Western Province. Since this field is an extension of the Moxico field of eastern Angola a similar (Aptian) age is inferred.

TANZANIA

Early Cretaceous strata in southern Tanzania occur in the upper part of the **Tendaguru Supergroup**. They are exposed on the Coastal Plateau to the east of Lindi and Kilwa where the succession rests disconformably on the Tithonian *smeei*-Oolite.

Along the northern rim of the Mandawa-Mahokondo anticline a thin (12-65 m thick) persistent, well-bedded white coralliferous limestone, the **Kikundi Formation**, is taken to be earliest Cretaceous. It has yielded colonial corals, stromatolites and occasional internal moulds of large gastropods. Along the Lihimalio Stream similar limestones contain the ammonites *Euphylloceras*, *Eulytoceras*,

Spitidiscus and *Barremites*, and the bivalves *Gervillella*, *Sphaera* and *Megacucullaea*, for which an Hauterivian age is indicated. At Mikadi the Early Hauterivian has yielded, in addition, *Olcostephanus*, *Jeannoticeras*, *Bornhardticeras* and *Phyllopachyceras*.

Conformably overlying the Kikundi is the **Namitambo Formation** comprising some 85 m of dark grey medium-grained sandstones with finer interbeds. It has yielded the bivalves *Venilicardia* and *Megacucullaea* of probable Barremian age.

The **Ntandi Formation** follows disconformably on the Namitambo, overstepping onto Late Jurassic strata. It comprises soft-weathering, grey-buff, fine- to medium-grained calcareous sandstones with fossiliferous gritty bands and pebble beds. At the base is the Barremian "**Ntandi fauna**" with the brachiopods *Zeilleria* and *Kingena*, the belemnite *Duvalia*, and the bivalves *Gervillella*, *Pseudoyaadia*, *Turikirella bornhardti* and *T. turikirae*, *Entolium*, *Camptonectes*, *Neithea atava*, *Hinnites*, *Aetostreon*, *Astarte*, *Eriphyla*, *Avicula* and *Ptychomya*.

The upper part of the Ntandi Formation is earliest Aptian. Besides the index species *Tanzanitrigonia schwarzi* it has yielded, in addition, *Zulutrigonia*, *Aetostreon*, *Astarte*, *Tancredia*, *Globocardium*, *Neithea*, *Megacucullaea*, *Pholadomya*, *Hinnites*, *Ptychomya*, *Plagiostoma*, *Sphaera* and *Stegoconcha*, the nautiloids *Cenoceras* and *Cymatoceras*, the echinoid *Pygurus*, and the ammonites *Ancyloceras*, *Procheloniceras* and *Kutattusites*.

The **Kiturika Formation** follows disconformably on gently-folded and faulted older Cretaceous strata, and is dated to the Late Aptian. Along the eastern and south-eastern flanks of the Ngarama Plateau it is a white reefal limestone and calcarenite 60-160 m thick. Stromatolites abound with occasional corals; *Orbitolina* is a common foraminifer.

The **Makonde Formation** is imprecisely dated and may represent a terrestrial redbed facies of the Kiturika. It comprises unfossiliferous red, purple and brown sandstones with laminated mudstones.

Albian strata overstep the Aptian and, in the vicinity of the Kikundi stream, comprise mainly green marls with thin interbeds of ripple-marked grey calcareous sandstones with trace fossils. Both the Early Albian with *Douvilleiceras* and the Late Albian with *Mortoniceras*, *Puzosia*, *Parasilesites* and *Euphylloceras* are present, but the precise stratigraphy is unknown.

In the Mbeya district of SW Tanzania, terrestrial redbeds similar to those of the Karonga Formation of Malawi, are exposed in the Rukwa Trough. The predominantly fluviatile succession, assigned to the **Galula Formation** (= "Red Sandstone Group"), rests unconformably on Karoo sediments and is a braidplain deposit of 600-3000 m of mainly red and pink, medium-grained, planar and trough cross-bedded sandstones with pebble lags and interbedded mudstones. The little-studied fauna includes titanosaurid and theropod dinosaurs, gondwanotherian mammals, teleost fishes, megaloolithid eggshells and a bird. A Late Cretaceous age is favoured.

Acknowledgements

This book has been 45 years in compilation, and many people along the way have provided the criticism and encouragement required for its completion. These include the late Dr S. H. Haughton (Pretoria), Prof. W. J. Verwoerd (Stellenbosch), the late Prof. J. de Villiers (Cape Town), the late

Dr B. F. Kensley (Cape Town and Washington), Prof. N. M. Savage (Oregon), Prof. W. J. Kennedy (Oxford), Prof. P. Bengtson (Heidelburg), Dr H. G. Owen (London), and last but not least my wife Brenda and my children Fern, David, Mike and Rose. This book is a compilation of the work of others whose contributions can be found in the reference list.

References

Aberhan, M., Bussert, R., Heinrich, W.-O. & Kapilima, S. 2002. Palaeoecology and depositional environment of the Tendaguru Beds (Late Jurassic to Early Cretaceous, Tanzania). *Mittel. Mus. Natur. Berlin, Geowissen.* **5**: 19–44.

Adamson, R. S. 1931. Note on some petrified wood from Banke, Namaqualand. *Trans. R. Soc. S. Afr.* **19**: 255–258.

—— 1934. Fossil plants from the Fort Grey area, East London. *Ann. S. Afr. Mus.* **31**: 67–96.

Alberti, A., Piccirillo, E. M., Bellieni, G., Civetta, L., Comin-Chiaramonti, P., Morais, E. A. A. 1992. Mesozoic acid volcanics from southern Angola:petrology, Sr-Nd isotope characteristics and correlation with the acid stratoids volcanic suites of the Parana basin (south-eastern Brazil). *Eur. J. Mineral.* **4**: 597–604.

Alberti, A., Castorina, F., Censi, P., Comin-chiaramonti, P., Gomes, C.B.,1999. Geochemical characteristics of Cretaceous carbonatites from Angola. *J. Afric. Earth Sci.* **29**(4): 735–759.

Alvarez, L. W., Alvarez, W., Asaro, F., Michel, H. V. 1980. Extraterrestrial cause for the Cretaceous Tertiary extinction. *Science, N.Y.* **208**:1095–1108.

Anderson, W. 1902. Report on a reconnaissance geological survey of the eastern half of Zululand, with a geological sketch of the country traversed. *Rep. geol. Surv. Natal Zululand* **1**: 37–66.

—— 1904. Further notes on the reconnaissance geological survey of Zululand. *Rep. geol. Surv. Natal Zululand* **2**: 39–67.

—— 1906. The Upper Cretaceous rocks of Natal & Zululand. *Rep. geol. Surv. Natal Zululand* **3**: 47–64.

Andrade, M. M. de. 1957. Rochas vulcânicas da orla Meso-Cenozóica entre Benguela e Moçâmedes. *Garçia de Orta* **5**: 739–766.

—— & Andrade, J.B. M. de. 1957. Estado actual dos conhecimentos sobre a paleontologia de Angola (até fins de 1955). *An. Estud. Geol. Paleont., Jta Inves. Ultramar* **12**(7): 209.

Antunes, M. T. 1961. Sur la faune de vertébrés du Crétacé de Iembe (Angola). *C. r. hebd. Séanc. Acad. Sci., Paris* **253**: 513–514.

—— 1964. O Neocretácico e o Cenozóico do litoral de Angola. *Jta Inves. Ultramar* **27**: 1–257.

—— & Cappetta, H. 2002. Sélaciens du Crétacé (Albien-Maastrichtien) de Angola. *Palaeontographica* (A) **264**: 85–146.

—— & Sornay, J. 1969. Contribution à la connaissance du Crétacé Superieur de Barra do Dande, Angola. *Rev. Fac. Cienc. Univ. Lisbon* (2, C) **16**: 65–104.

Araújo, R., Polcyn, M., Mateus, O. & Schulp, A. 2010. Plesiosaurs from the Maastrichtian of Bentiaba, Namibe Province, Angola. *J. vert. Paleont.* **30**(3): 55A.

Atherstone, W. G. 1857. Geology of Uitenhage. *E. Prov. Monthly Mag.* **1**: 518–532, 579–595.

Baily, W. H. 1855. Description of some Cretaceous fossils from South Africa; collected by Capt. Garden, of the 45th Regiment. *Jl geol. Soc. Lond.* **11**: 454–465.

Bandel, K. & Kiel, S. 2003a Relationships of Cretaceous Neritimorpha (Gastropoda, Mollusca), with the description of seven new species. *Bull. Czech. geol. Surv.* **78**(1): 53–65.

Bandel, K. & Kiel, S. 2003b. New taxonomic data for the gastropod fauna of the Umzamba Formation (Santonian–Campanian, South Africa) based on newly collected material. *Cret. Res.* **24**: 449–475.

BASSE, E. 1963. Quelques ammonites nouvelles du Crétacé Supérieur d'Angola. *Bull. Soc. géol. Fr.* (7) **4**: 871–876.

BATE, R. H. & BAYLISS, D. D. 1969. An outline account of the Cretaceous and Teriary Foraminifera and of the Cretaceous ostracods of Tanzania. *In*: Proc. 3rd Afr. Micropal. Colloq., Cairo:113–164.

BESAIRIE, H. 1930. Les rapports du Cretace malgache avec le Cretace de l'Afrique australe. *Bull. Soc. geol. Fr.* **30**(4): 613–643.

—— & LAMBERT, J. 1930. Note sur quelques echinides de Madagascar et du Zululand. *Bull. Soc. geol. Fr.* **30**(4): 107–117.

BLAKE, D. B., BRETON, G. & GOFAS, S. 1996. A new genus and species of Asteriidae (Asteroidea; Echinodermata) from the Upper Cretaceous (Coniacian) of Angola, Africa. *Paläont. Zeitschrift* **70**(1–2): 181–187.

BÖHM, J. & RIEDEL, L. 1933. Uber eine als *Placenticeras* beschriebene *Oppelia* (*Bornhardticeras* n.g.) aus dem Neokom des Deutsch-Ostafrika. *Jb. Preuss. geol. Land.* **53**: 112–124.

BOND, G. 1973. The palaeontology of Rhodesia. *Bull. geol. Surv. Rhod.* **70**: 1–121.

BORGES, A. 1944. Depósitos conglomeráticos do Alto Limpopo. *Bolm Servs. Indust. Minas Geol., Lourenço Marques* **6**: 5–12.

—— 1946. A costa de Angola da Baía da Lucira à foz do Bentiaba (Entre Benguela e Mossâmedes). *Bolm Soc. geol. Port.* **5**(3): 141–150.

BOSHOFF, J. C. 1945. *Stratigraphy of the Cretaceous System in the Nduma area of northern Zululand.* Unpubl. M. Sc. thesis, Univ. Pretoria.

BRENNER, P. W. & OERTLI, H. J. 1976. Lower Cretaceous ostracodes (Valanginian to Hauterivian) from the Sundays River Formation, Algoa basin, South Africa. *Bull. Centre Res. Pau, Soc. Nat. Petrol.Aquitaine* **10**: 471–533.

BRISTOW, J. W. & ALLSOPP, H.L. 1985. Summary of the geochronology of the southern Lebombo. *Spec. Publ. geol. Soc. S. Afr., Dept Geol., Univ. Natal (Durban)* **1**: 12–14.

—— & DUNCAN, A. R. 1983. Rhyolitic dome formation and plinian activity in the Bumbeni Complex, southern Lebombo. *Trans. geol. Soc. S. Afr.* **86**(3): 273–280.

—— & SAGGERSON, E. P. 1983. A review of Karoo vulcanicity in southern Africa. *Bull. Volcanol.* **46**: 135–159.

BROGNON, G. 1971. The geology of the Angola coast and continental margin. *Rep. Inst. Geol. Sci.* **70**(16): 147–152.

—— & VERRIER, G. R. 1955. Contribution à la géologie du bassin du Cuanza en Angola. *4th World Petrol. Congr., Rome,* Sect. **1**: 251–265.

—— & —— 1958. Note sur la stratigraphie du bassin Cuanza en Angola. *Bolm Soc. géol. Port.* **12**: 61–74.

—— & —— 1966. Oil and geology in the Cuanza basin of Angola. *Bull. Amer. Assoc. petrol. Geol.* **50**: 108–158.

BROMLEY, R. G. 1974. Trace fossils at omission surfaces. *In*, FREY, R. W., ed., *The study of trace fossils*. Springer-Verlag: Berlin.

BROOD, K. 1977. Upper Cretaceous Bryozoa from Need's Camp, South Africa. *Palaeont. afr.* **20**: 65–82.

BROOM, R. 1904. On the occurrence of an opisthocoelian dinosaur (*Algosaurus bauri*) in the Cretaceous beds of South Africa. *Geol. Mag.* **5**: 445–477.

—— 1907. On some reptilian remains from the Cretaceous beds at the mouth of the Umpenyati River. *Rep. geol. Surv. Natal Zululand* **3**: 95.

—— 1912. On a species of *Tylosaurus* from the Upper Cretaceous beds of Pondoland. *Ann. S. Afr. Mus.* **7**: 332–333.

BROWN, J. T. 1977a. The morphology and taxonomy of *Cycadolepis jenkinsiana* and *Zamites recta* from the Lower Cretaceous Kirkwood Formation of South Africa. *Palaeont. afr.* **20**: 43–46.

—— 1977b. On *Araucarites rogersii* Seward for the Lower Cretaceous Kirkwood Formation of the Algoa basin, Cape Province, South Africa. *Palaeont. afr.* **20**: 47–51.

BROWNFIELD, M. E.& CHARPENTIER, R. R. 2006. Geology and total petroleum systems of the west-central coastal province (7203), West Africa. *Bull. U. S. geol. Surv.* **2207–B**: 1–52.

BUSSERT, R., HEINRICH, W. D. & ABERHAN, A. 2009. TheTendaguru Formation (Late Jurassic to Early Cretaceous, southern Tanzania): definition, palaeoenvironments, and sequence stratigraphy. *Fossil Record* **12**(2): 141–174.

CALZADA, S. & CORBACHO, J. 2014. Sobre dos Nerineas de Angola. *Battaleria* **20**: 16–20.

CARVALHO, G. S. DE. 1958. As formações cretácicas da bacia de Moçâmedes (Angola) e alguns dos sevs problemas. *Publ. Mus. Lab. min. geol. Porto* **75**(3): 1–32.

—— 1960. Sobre os depósitos Cretácicos do litoral de Angola. *Bolm Serv. geol. Min. Angola* **1**: 37–48.

—— 1961. Geologia do deserto de Moçâmedes (Angola). *Mems Jta Invest. Ultramar* **26**(2): 1–227.

—— 1968. Some problems concerning the mineral occurrences in the Lower Cretaceous of the Moçâmedes sedimentary basin (Angola). *Garcia de Orta, Rev. Junta Miss. Geogr. Invest.Ultramar* **16**(1): 93–106.

CASIER, E. 1957. Les faunes icthyologiques du Crétacé et du Cénozoïque de l'Angola et de l'enclave de Cabinda. Leurs affinités paléobiogéographiques. *Comunções Servs geol. Port.* **38**(2): 1–267.

—— 1961. Matériaux pour la faune ichthyologique. Eocrétacique du Congo. *Ann. Mus. Roy. Afr. Centr.*, Série 8, Sci. Géol. **39**: 1–96.

CHOFFAT, P. 1887a. Note préliminaire sur des fossiles recueillis par M. Lourenço Malheiro, dans la province de l'Angola. *Bull. Soc. Geol. Fr.*, 3 serie, **15**: 154–157.

—— 1887b. Kreideablagerungen an der WestKüste von Süd Afrika. *N. Jb. Min. Geol. Paläont.* **1**: 117–118.

—— 1905. Nouvelles données sur la zone littorale d'Angola. *Comunções Servs geol. Port.* **1905**: 31–78.

—— & LORIOL, P. DE. 1888. Matériaux pour l'étude stratigraphique et paléontologique de la province d'Angola. *Mem. Soc. Phys. Geneve* **30**: 1–116.

COOPER, M. R. 1972. The Cretaceous stratigraphy of São Nicolau and Salinas, Angola. *Ann. S. Afr. Mus.* **60**(8): 245–251.

—— 1973. Cenomanian ammonites from Novo Redondo, Angola. *Ann. S. Afr. Mus.* **62**(2): 41–67.

—— 1974. The Cretaceous stratigraphy of south-central Africa. *Ann. S. Afr. Mus.* **66**: 81–107.

—— 1976. The mid-Cretaceous (Albian-Turonian) biostratigraphy of Angola. *Ann. Mus. Hist. nat. Nice* **4**: xvi.1–20.

—— 1977. Eustacy during the Cretaceous: its implications and importance. *Palaeogeogr., Palaeoclima., Palaeoecol.* **22**: 1–60.

—— 1981. Revision of the Late Valanginian Cephalopoda from the Sundays River Formation of South Africa, with special reference to the genus *Olcostephanus*. *Ann. S. Afr. Mus.* **83**: 147–366.

—— 1988. A new species of trigoniid bivalve from the Lower Cretaceous of Zululand. *S. Afr. J. Geol.* **91**(3): 326–328.

—— 1989a. A new species of *Linotrigonia* (Mollusca: Bivalvia) from the Campanian of Zululand. *Palaeont. afr.* **26**(9): 99–103.

—— 1989b. The Gondwanic bivalve *Pisotrigonia* (family Trigoniidae), with description of a new species. *Paläont. Zeitschrift* **63**(3/4): 241–250.

—— 1990. A new genus of Rutitrigoniinae (Bivalvia, Trigoniacea) from the Lower Cretaceous (Aptian) of Zululand. *Ann. S. Afr. Mus.* **99**(3): 23–29.

—— 1991. Lower Cretaceous Trigonioida (Mollusca, Bivalvia) from the Algoa basin, with a revised classification of the order. *Ann. S. Afr. Mus.* **100**(1): 1–52.

—— 1993. A new species of *Megatrigonia* (Bivalvia: Trigonioida) from northern Zululand, with comments on the phylogeny of the genus. *Durban Mus. Novit.* **18**: 21–28.

—— 1996. Tectonic cycles in Southern Africa. *Earth-Sci. Rev.* **28**: 321–364.

—— 1997. Exogyrid oysters (Bivalvia: Gryphaeoidea) from the Cretaceous of southeast Africa. Part 2. *Durban Mus. Novit.* **22**: 1–31.

—— 2000. The first record of the Gondwanic bivalve *Nototrigonia* (Trigonioida: Neotrigoniidae) from the Lower Cretaceous of South Africa, with description of a new species. *Durban Mus. Novit.* **25**: 1–4.

—— 2015a. On the Pterotrigoniidae van Hoepen, 1929 (Bivalvia:Trigoniida); their biogeography, evolution, classification and relationships. *N. Jb. Geol. Paläont. Abh.* **277**(1): 11–42.

—— 2015b. On the Iotrigoniidae van Hoepen, 1929 (Bivalvia:Trigoniida); their palaeobiogeography, evolution and classification. *N. Jb. Geol. Paläont. Abh.* **277**(1): 49–62.

—— 2015c. On the Rutitrigoniidae van Hoepen, 1929 (Bivalvia:Trigoniida); their palaeobiogeography, evolution and classification. *N. Jb. Geol. Paläont. Abh.* **278**(2): 159–173.

—— 2016. On the Pleurotrigoniinae van Hoepen, 1929 (Trigoniida: Trigoniidae); their evolution and relationships. *N. Jb. Geol. Paläont. Abh.* **279**(1): 23–34.

—— & Leanza, H. A.. 2017. On the Steinmanellidae (Bivalvia: Myophorelloidea); their palaeobiogeography, evolution and classification. *N. Jb. Geol. Paläont. Abh.* **285**(3): 313–325.

Cox, K.1963. Malvernia Beds. *Trans. Proc. geol. Soc. S. Afr.* **66**: 341–344.

Cox, L. R. 1925. Cretaceous Gastropoda from Portuguese East Africa. *Ann. Transv. Mus.* **11**: 201–216.

Chapman, F. 1904. Foraminifera and Ostracoda from the Cretaceous of east Pondoland, South Africa. *Ann. S. Afr. Mus.* **4**: 221–237.

—— 1916. Foraminifera and Ostracoda from the Upper Cretaceous of Needs Camp, Buffalo River, Cape Province. *Ann. S. Afr. Mus.* **12**: 107–118.

—— 1923. On some Foraminifera and Ostracoda from the Cretaceous of Umzamba River, Pondoland. *Trans. Proc. geol. Soc. S. Afr.* **28**: 1–6.

Cracraft, J. 1973. Continental drift, paleoclimatology, and the evolution and biogeography of birds. *Jl Zool. Lond.* **169**: 455–545.

Crick, G. C. 1907a. The Cephalopoda from the deposit at the north end of False Bay, Zululand. *Rep. geol. Surv. Natal Zululand* **3**: 163–234.

—— 1907b. The Cephalopoda from the tributaries of the Manuan Creek, Zululand. *Rep. geol. Surv. Natal Zululand* **3**: 235–249.

—— 1907c. Note on a Cretaceous ammonite from the mouth of the Umpenyati River, Natal. *Rep. geol. Surv. Natal Zululand* **3**: 250.

—— 1907d. The Cretaceous rocks of Natal and Zululand and their cephalopod fauna. *Geol. Mag.* **4**: 339–347.

—— 1924. On Upper Cretaceous Cephalopoda from Portuguese East Africa. *Trans. Proc. geol. Soc. S. Afr.* **26**: 130–140.

Dacqué, E. & Krenkel, E. 1909. Jura und Kreide in Ostafrika. *Neues Jb. Geol. Min. Paläont.* Beil-Bd **27**: 150–232.

Dartevelle, E. 1942. Le Crétacé Supérieur de Mossamedes (contribution à la geologie de l'Angola. *Bull. Soc. belge Géol. Paléont. Hydrol.* **50**: 186–189.

—— 1952. Echinides fossiles du Congo et de l'Angola. I. Introduction historique et stratigraphique. *Ann. Mus. Congo belge 8vo* **12**: 1–71.

—— 1953. Echinides fossiles du Congo et de l'Angola. I. Description systematique des echinides fossiles du Congo et de l'Angola. *Ann. Mus. Congo belge 8vo* **13**: 1–240.

—— & Brebion, P. 1956. Mollusques fossiles du Crétacé de la côte occidentale d'Afrique du Cameroun à l'Angola. I. Gasteropodes. *Ann. Mus. Congo belge 8vo* **15**: 1–128.

—— & CASIER, E. 1941. Les poissons fossiles de l'Angola. *Comunções Servs geol. Port.* **22**: 99–109.

—— & —— 1946. Les poissons fossiles de l' Angola, *Comunções Servs geol. Port.* **27**: 85–90.

—— & FRENEIX, S. 1957. Mollusques fossiles du Crétacé de la côte occidentale d'Afrique du Cameroun à l'Angola, *Ann. Mus. Congo belge 8vo* **20**: 1–272.

DAVEY, R. J. 1969. Some dinoflagellate cysts from the Upper Cretaceous of northern Natal, South Africa. *Palaeont. afr.* **12**: 1–15.

DE KLERK, W. J., FORSTER, C. A., ROSS, C. F., SAMPSON, C. D. & CHINSAMY, A. 1997. New maniraptoran and iguanodontian dinosaurs from the Early Cretaceous Kirkwood Formation, South Africa. *J. Vert. Paleont.* **17**(Suppl.): 42A.

——, ——, ——, —— & —— 1998. A review of recent dinosaur and other vertebrate discoveries in the Early Cretaceous Kirkwood Formation in the Algoa Basin, Eastern Cape, South Africa. *J. Afr. Earth Sci.* **27**: 55.

——, ——, SAMPSON, S. D., CHINSAMY, A. & ROSS, C. F. 2000. A new coelurosaurian dinosaur from the Early Cretaceous of South Africa. *J. Vert. Paleont.* **20**(2): 324–332.

DE WIT, M. C. J., WARD, J. D., BAMFORD, M. K. & ROBERTS, M. J. 2009. The significance of the diamondiferous gravel deposits at Mahura Muthla, Northern Cape Province, South Africa. *S. Afr. J. Geol.* **112**: 89–108.

DINGLE, R. V. 1968a. Marine Neocomian Ostracoda from South Africa. *Trans. r. Soc. S. Afr.* **38**: 139–164.

—— 1968b. Upper Senonian ostracods from the coast of Pondoland, South Africa. *Trans. r. Soc. S. Afr.* **38**: 347–386.

—— 1969. Marine Neocomian Ostracoda from South Africa. *Trans. r. Soc. S. Afr.* **38**: 139–164.

—— 1971. Some Cretaceous ostracod assemblages from the Aghulas Bank (South African continental margin). *Trans. r. Soc. S. Afr.* **39**: 393–418.

—— 1973. Mesozoic palaeogeography of the southern Cape, South Africa. *Palaeogeogr., Palaeoclima., Palaeoecol.* **13**: 203–213.

—— 1980. Marine Santonian and Campanian ostracods from a borehole at Richards Bay, Zululand. *Ann. S. Afr. Mus.* **82**: 1–70.

—— 1981. The Campanian and Maastrichtian Ostracoda of south-east Africa. *Ann. S. Afr. Mus.* **85**: 1–181.

—— 1982.Some aspects of Cretaceous ostracod biostratigraphy of South Africa and relationships with other Gondwanide localities. *Cret. Res.* **3**: 367–389.

—— & KLINGER, H. C. 1971. Significance of Upper Jurassic sediments in the Knysna outlier (Cape Province) for timing of the breakup of Gondwanaland. *Nature (Phys. Sci.)* **232**: 37–38.

—— & —— 1972. The stratigraphy and Ostracoda fauna of the Upper Jurassic sediments from Brenton, in the Knysna outlier, Cape Province. *Trans. r. Soc. S. Afr.* **40**: 279–298.

——, SIESSER, W. G. & NEWTON, A. R. 1983. *Mesozoic and Tertiary geology of Southern Africa*. A. A. Balkema: Rottendam.

DIEITRICH, W. O. 1914. Die Gastropoden der Tendaguruschichten, der Aptstufe und der Oberkreide im südliche Deutsch-Ostafrika. *Arch. Biontol.* **3**(4): 97–153.

—— 1926. Steinkorallen des Malms und der Unter-kreide im südliche Deutsch-Ostafrika. *Palaeontographica* Suppl. 7, II Reihe, Teil I, Lfg **2**: 41–102.

——, 1933. Zur Stratigraphie und Paläontologie der Tendaguruschichten. *Palaeontographica* **7**: 1–86.

—— 1938. Zur Stratigraphie der Kreide im nordlichen Zululand. *Zentr. Miner Geol, Paläont* (B) **1938**: 228–240.

DIXEY, F. 1928. The Dinosaur Beds of Lake Nyasa. *Trans. r. Soc. S. Afr.* **16**: 55–66.

—— 1945. Fossils from the Pipe Sandstone at Victoria Falls, Rhodesia. *Trans. Proc. geol. Soc. S. Afr.* **47**: 5–7.

—— & CAMPBELL-SMITH, W. 1929. The rocks of the Lupata Gorge and the north side of the lower Zambesi. *Geol. Mag.* **66**: 241–259.

DOUVILLÉ, H. 1931. Contribution à la géologie de l'Angola. Les ammonites de Salinas. *Bolm Mus. Lab. miner. géol. Univ. Lisbon* **1**: 17–46.

ERASMUS, T. 1976. A new species of *Dammaroxylon* Schultze-Motel, *D. natalense* sp. nov., from the Cretaceous of Natal, South Africa. *Palaeont. afr.* **19**: 135–139.

ERNST, G. & ZANDER, J. 1993. Stratigraphy, facies development, and trace fossils of the Upper Cretaceous of southern Tanzania (Kilwa District). *In*: Geology and mineral resources of Somalia and surrounding areas. *Inst. Agron. Oltremare Firenze, Relaz. E Monogr.* **113**: 259–278.

ESTES, R. 1977. Relationship of the South African fossil frog *Eoxenopoides reuningi* (Anura, Pipidae). *Ann. S. Afr. Mus.* **73**: 49–80.

ETHERIDGE, R. 1904. Cretaceous fossils of Natal. I. The Umkwelane Hill deposit. *Rep. geol. Surv. Natal Zululand* **2**: 71–93.

—— 1907. Cretaceous fossils of Natal. II. The Umzinene River deposit. *Rep. geol. Surv. Natal Zululand* **3**: 67–90.

FAHRION, H. 1937. Die Foraminiferen der Kreide und Tertiär-Schichten im súdlichen Deutsch-Ostafrika. *Palaeontographica*, Suppl. VII, Zweite Reihe, **II**: 187–216.

FERRÉ, B. & GRANIER, B. 2001. Albian roveacrinids from the southern Congo Basin off Angola. *Jl S. Amer. Sci.* **14**: 219–235.

FERREIRA, O. 1957. Acerca de "*Parapirimela angolensis*" Van Straelen nas Camadas de Iela (Angola). *Com. Servs geol. Port.* **38**: 465–468.

FLORES, G. 1964. On the age of the Lupata rocks, lower Zambesi valley, Mozambique. *Trans. Proc. geol. Soc. S. Afr.* **67**: 111–118.

—— 1967. The Cretaceous and Tertiary sedimentary basins of Mozambique and Zululand. *In*, BLANT, G., ed.. *Sedimentary basins of the African coasts. 2nd Part, South and East Coasts*: 81–111. Assoc. Afr. geol. Survs: Paris.

FORSTER, C. A., FROST, S. & ROSS, C. F. 1995. New dinosaur material and paleoenvironment of the Early Cretaceous Kirkwood Formation, Algoa Basin, South Africa. *J. Vert. Paleont.* **15**(Suppl.): 29A.

FÖRSTER, R. 1975. Die geologische Entwicklung von Süd-Mozambique seit die Unterkreide und die Ammoniten-Fauna von Unterkreide und Cenoman. *Geol. Jahrb.* (B) **12**: 1–284.

FRANKEL, J. J. 1961. The geology along the Umfolosi River, south of Matubatuba, Zululand. *Trans. Proc. geol. Soc. S. Afr.* **63**: 231–252.

FRENEIX, S. 1959. Mollusques fossiles du Crétacé de la côte occidentale d'Afrique, du Cameroun a l'Angola. *Ann. Mus. Congo belge 8vo* **24**: 1–126.

GALTON, P. M. & COOMBS, W. P. 1981. *Paranthodon africanus* (Broom) a stegosaurian dinosaur from the Lower Cretaceous of South Africa. *Geobios* **14**: 299–309.

GARDEN, R. J. 1855. Notice of some Cretaceous rocks near Natal, South Africa. *Quart. Jl geol. Soc. Lond.* **11**: 453–454.

GEVERS, T. W. & LITTLE, J. DE V. 1946. Upper Cretaceous beds between the Intongazi and Umkandandhlovu Rivers, Alfred County, Natal. *Trans. Proc. geol. Soc. S. Afr.* **48**: 27–29.

GIERLOWSKI-KORDESCH, E. & ERNST, G. 1987. A flysch trace assemblage from the Upper Cretaceous shelf of Tanzania. *In*: Matheis, G., Schandelmeier, H. (eds.), *Current research in African earth science*: 217–222. Rotterdam.

GOMANI, E. M. 2005. Sauropod dinosaurs from the Early Cretaceous of Malawi, Africa. *Palaeontologia Electronica* **8**(1): 1–36.

GOTTSCHE, C. 1887. Ueber die obere Kreide von Umtafuna (S. Natal). *Z. dt geol Gesell* **39**: 622–624.

GREGORY, J. W. 1916. Contributions to the geology of Benguella. *Trans. Roy. Soc. Edinb.* **51**(3): 495–536.

—— 1930. A new *Ceratotrochus* from the Upper Cretaceous of Portuguese East Africa. *Geol. Mag.* **67**: 475–477.

GREYLING, E. H. & COOPER, M. R. 1993. A new cassiduloid echinoid from the Lower Cretaceous (Upper Albian) of Zululand. *Durban Mus. Novit.* **18**: 13–20.

—— & —— 1994. Little-known irregular echinoids from the Lower Cretaceous (Upper Albian) of Zululand. *Durban Mus. Novit.* **19**: 41–58.

—— & —— 1995. Two new irregular echinoids from the Upper Cretaceous (Campanian) of Angola. *Durban Mus. Novit.* **20**: 63–71.

GRIESBACH, C. L. 1871. On the geology of Natal in South Africa. *Quart. Jl geol. Soc. Lond.* **27**: 53–72.

HAAS, O. 1942a. The Vernay Collection of Cretaceous (Albian) ammonites from Angola. *Bull. Am. Mus. nat. Hist.* **81**: 1–224.

—— 1942b. Some Upper Cretaceous ammonites from Angola. *Am. Mus. Novit.* **1182**: 1–24.

—— 1943. Some abnormally coiled ammonites from the Upper Cretaceous of Angola. *Am. Mus. Novit.* **1222**: 1–17.

—— 1945. A recently acquired Albian ammonite from Angola. *Am. Mus. Novit.* **1286**: 1–4.

—— 1952. Some Albian desmoceratid and lytoceratid ammonites from Angola. *Am. Mus. Novit.* **1561**: 1–17.

HAUGHTON, S. H. 1924. Note sur quelques fossiles Crétacé de l'Angola (céphalopodes et echinides). *Comunçoes Servs. geol. Port.* **15**: 79–106.

—— 1925. Notes on some Cretaceous fossils (Cephalopoa and Echinoidea) from Angola. *Ann. S. Afr. Mus.* **22**: 263–288.

—— 1930a. Note on the occurrence of Upper Cretaceous marine beds in South West Africa. *Trans. Proc. geol. Soc. S. Afr.* **33**: 61–63.

—— 1930b. On the occurrence of Upper Cretaceous marine fossils near Bogenfels, S. W. Africa. *Trans. r. Roc. S. Afr.* **18**: 361–365.

—— 1931. On a collection of fossil frogs from the clays at Banke. *Trans. r. Soc. S. Afr.* **19**: 233–249.

—— & BOSHOFF, J. C. 1956. Algumas ammonites Aptianas de Chalala (Africa Oriental Portuguesa). *Servs Ind. geol. Moç.* (Geol.) **117**: 1–24.

——, FROMMURZE, F. H. & & VISSER, D. J. L. 1937. The geology of portions of the coastal belt near the Gamtoos valley, Cape Province. *Explan. Sheets 151 geol. Surv. S. Afr.*: 1–55.

—— 1963. *The stratigraphic history of Africa south of the Sahara.* Oliver & Boyd: Edinburgh.

HENNIG, E. 1914. Beitrage zur Geologie und Stratigraphie Deutsch-Ostafrika. I. Geologisch-stratigraphische Beobachtungen im Kustengebiet des südlichen Deutsch-Ost-afrika. II. Geologisch-stratigraphische Beobachtungen im Gebiete der Jura-Ablager-ungen an der Deutsch-Ostafrikanischen Zentralbahn. *Arch. Biontol.* **3**(3): 1.

—— 1916. Die Fauna der Deutsch-ostafrikanischen Urgonfazies. *Zt. dtsch. geol. Ges.* **68**: 441.

—— 1937. Der Sedimentstreifen des Lindi-Kilwa Hinterlandes (Deutsch-Ostafrika). *Palaeontographica*, Suppl. 7, Reihe II, Lfg **2**: 99.

HEINZ, R. 1930. Ueber Kreide-Inoceramen der sud-afrikanischen Union. *C. r. XV Int. geol. Congress* **2**: 681–687.

HODGSON, F. D. I. & BOTHA, B. J. V. 1976. The Karroo sediments in the vicinity of Doros, South West Africa. *Ann. geol. Surv. S. Afr.* **10**: 49–56.

HOEPEN, E. C. N. VAN. 1920. Description of some Cretaceous ammonites from Pondoland. *Ann. Transvaal Mus.* **7**: 142–147.

—— 1921. Cretaceous Cephalopoda from Pondoland. *Ann Transv Mus.* **8**: 1–48.

—— 1929. Die Krytfauna van Soeloeland. I. Trigoniidae. *Palaeont Narvors nas Mus Bloemfontein* **1**: 1–38.

—— 1931. Die krytfauna van Soeloeland. 2. Voorlopige beskrywing van enige Soeloelandse ammoniete (i). *Lophoceras, Rhytidoceras, Drepanoceras* en *Deiradoceras.*

Palaeont. Navors. nas. Mus. Bloemfontein **1**(2): 37–54.

—— 1941. Die gekielde ammoniete van die Suid-Afrikaanse Gault. I. Dipoloceratidae, Cechenoceratidae en Drepanoceratidae. *Palaeont. Navors. nas. Mus. Bloemfontein* **1**(3): 55–90.

—— 1942. Die gekielde ammoniete van die Suid-Afrikaanse Gault. II. Drepanoceratidae, Pervinquieridae, Arestoceratidae, Cainoceratidae. *Palaeont. Navors. nas. Mus. Bloemfontein* **1**(4): 91–157.

—— 1944. Die gekielde ammoniete van die Suid-Afrikaanse Gault. III. Pervinquieridae en Brancoceratidae. *Palaeont. Navors. nas. Mus. Bloemfontein* **1**(5): 159–198.

—— 1946a. Die gekielde ammoniete van die Suid-Afrikaanse Gault. IV. Cechenoceratidae, Dipoloceratidae, Drepanoceratidae, Aresto-ceratidae. *Palaeont. Navors. nas. Mus. Bloem-fontein* **1**(6): 199–260.

—— 1946b. Die gekielde ammoniete van die Suid-Afrikaanse Gault. V. Monophyletism or polyphyletism in connection with ammonites of the South African Gault. *Palaeont. Navors. nas. Mus. Bloemfontein* **1**(7): 261–271.

—— 1951a. Die gekielde ammoniete van die Suid-Afrikaanse Gault. VI. The so-called old mouth-edges of the ammonites shell. *Palaeont. Navors. nas. Mus. Bloemfontein* **1**(8): 273–284.

—— 1951b. Die gekielde ammoniete van die Suid-Afrikaanse Gault. VII. Pervinquieridae, Arestoceratidae, Cainoceratidae. *Palaeont. Navors. nas. Mus. Bloemfontein* **1**(9): 285–344, 388–422.

—— 1951c. A remarkable desmoceratid from the South African Albian. *Palaeont. Navors. nas. Mus. Bloemfontein* **1**(10): 345–349.

—— 1955a. New and little-known ammonites from the Albian of Zululand. *S. Afr. J. Sci.* **51**: 355–361.

—— 1955b. Turonian-Coniacian ammonites from Zululand. *S. Afr. J. Sci.* **51**: 361–377.

—— 1955c. A new family of keeled ammonites from the Albian of Zululand. *S. Afr. J. Sci.* **51**: 377–382.

—— 1957. The deposits on the Umsinene River. *Compte Ren. CCTA, Conf. géol., Tananarive* **2**: 349–350.

—— 1964. An Albian astracurid from Zululand. *Ann. geol. Surv. S. Afr.* **1**: 253–255.

—— 1965. The Peroniceratinae and allied forms of Zululand, South Africa. *Mem. geol. Surv. S. Afr.* **55**: 1–70.

—— 1968a. New and little known Zululand and Pondoland ammonites. *Ann. geol. Surv. S. Afr.* **4**(1965): 158–181.

—— 1968b. New ammonites from Zululand. *Ann. geol. Surv. S. Afr.* **4**(1965): 183–191.

Hoppener, H. 1958. Brief report on the palaeontology of the Cuanza basin. *Bolm Soc. geol. Port.* **12**: 75–82.

Howarth, M. K. 1965. Cretaceous ammonites and nautiloids from Angola. *Bull. Br. Mus. nat. Hist.* (Geol.) **10**: 335–412.

—— 1968. A mid-Turonian ammonite fauna from the Moçâmedes desert. *Garçia de Orta* **14**: 217–228.

—— 1985. Cenomanian and Turonian ammonites from the Novo Redondo area. *Bull. Br. Mus. nat. Hist.* (Geol.) **39**(2): 73–105.

Janensch, W. & Hennig, E. 1914. Tabellarische Ubersicht des Fundforte wirbellose Fossilien im Arbeitsgebiete der Tendaguru-Expedition. *Arch. Biontol.* **3**(4): 1.

Jacobs, L. L., Mateus, O., Polcyn, M. J., Schulp, A. S., Antunes, M. T., Morais, M. L. & Tavares, T. 2006a. The occurrence and geological setting of Cretaceous dinosaurs, mosasaurs, plesiosaurs, and turtles from Angola. *J. paleont. Soc. Korea* **22**(1): 91–110.

——, —— , ——, ——, Scotese, C. R., Goswami, A., Ferguson, K. M., Robbins, J. A., Vineyard, D. P. & Neto, A. B. 2009a. Cretaceous paleogeography, paleoclimatology, and amniote biogeography of the low and mid-latitude South Atlantic Ocean. *Bull. Soc. géol. Fr.* **180**(4): 239–247.

——, Morais, M. L., Schulp, A. S., Mateus, O. & Polcyn, M. J. 2006b. Systematic position and geological context of *Angolasaurus* (Mosasauridae) and a new sea turtle from the Cretaceous of Angola. *J. vert. Paleont.* **26**(3): 81A.

—— , Polcyn, M., Araújo, R., Strganac, C. and Mateus, O. 2010. Physical drivers of evolution and the history of the marine tetrapod fauna of Angola. *J. vert. Paleont.* **30**(3): 110A.

——, Polcyn, M.J., Mateus, O., Schulp, A. & Neto, A.B. 2009. The Cretaceous Skeleton Coast of Angola. *J. vert. Paleont.* **29**(3): 121A.

——, ——, ——, —— , Ferguson, K., Scotese, C., Jacobs, B. F., Strganac, C., Vineyard, D., Myers, T. S., Morais, M. L. 2010. Tectonic drift, climate, and paleoenvironment of Angola since the Cretaceous. *Am.geophy. Union, Fall Meeting Abstr.*: 1–2.

——, Winkler, D. A. & Downs, W. R. 1992. Malawi's paleontological heritage. *Occ. Pap. Malawi Dept Antiquities* **1**: 5–22.

—— , —— & —— 1993. New material of an Early Cretaceous titanosaurid sauropod dinosaur from Malawi. *Palaeontology* **36**: 523–534.

—— , ——, Kaufulu, Z. M. & Downs, W. R. 1990. The Dinosaur Beds of northern Malawi, Africa. *Nat. Geogr. Res.* **6**: 196–204.

Jeletzky, J. A. 1983. Macroinvertebrate paleontology, biochronology and paleo-environments of Lower Cretaceous and Upper Jurassic rocks, Deep Sea Drilling Hole 511, eastern Falkland Plateau. *In*, Ludwig, W. J. *et al.*, eds., *Initial Reports of the Deep Sea Drilling Project* **71**: 951–975.

Jenkyns, H. C. 1971. The genesis of condensed sequences in the Tethyan Jurassic. *Lethaia* **4**: 327–352.

Kapilima, S. 2004. Tectonic and sedimentary evolution of the coastal basin of Tanzania during the Mesozoic times. *Tanz. J. Sci.* **29**(2003): 1–16.

Karpeta, W.R. 1986. The Cretaceous Mbotyi and Mngazana Formations of the Transkei coast: their sedimentology and structural setting. *S. Afr. J. Geol.* **90**: 25–36.

Kennedy, W. J. & Cooper, M. R. 1975. Cretaceous ammonite distributions and the opening of the South Atlantic. *Jl geol. Soc. Lond.* **131**: 283–288.

—— & Klinger, H. C. 1971. A major intra-Cretaceous unconformity in eastern South Africa. *Jl geol. Soc. Lond.* **127**: 183–186.

—— & —— 1972. Hiatus concretions and hardground horizons in the Cretaceous of Zululand, South Africa. *Palaeontology* **15**: 539–549.

—— & —— 1972. A *Texanites-Spinaptychus* association from the Upper Cretaceous of Zululand. *Palaeontology* **15**: 394–399.

—— —— 1977a. Cretaceous faunas from Zululand and Natal, South Africa: The ammonite family Tetragonitidae Hyatt, 1900. *Ann. S. Afr. Mus.* **73**: 149–197.

—— & —— 1977b. Cretaceous faunas from Zululand and Natal, South Africa: a *Jauberticeras* from the Mzinene Formation (Albian). *Ann. S. Afr. Mus.* **74**: 1–12.

—— & —— 1978. Cretaceous faunas from Zululand and Natal, South Africa: a new genus and species of Gastroplitinae from the Mzinene Formation (Albian). *Ann. S. Afr. Mus.* **77**: 57–69.

—— & —— 1979a. Cretaceous faunas from Zululand and Natal, South Africa: The ammonite superfamily Haplocerataceae Zittel, 1884. *Ann. S. Afr. Mus.* **77**: 85–121.

—— & —— 1979b. Cretaceous faunas from Zululand and Natal, South Africa. The ammonite family Gaudryceratidae. *Bull. Br. Mus. (Nat. Hist.)*, Geol. **31**(2): 121–174.

—— & —— 1993. On the affinities of *Zuluscaphites* van Hoepen 1955 (Cretaceous Ammonoidea) from the Albian of Zululand, South Africa. *Paläont. Zeitschrift* **67**: 63–67.

—— & —— 2008. Cretaceous faunas from Zululand and Natal, South Africa. The ammonite subfamily Lyelliceratinae Spath, 1921. *Afr. Nat. Hist.* **4**: 57–111.

—— & —— 2009. The heteromorph ammonite *Ndumuiceras variabile* gen. et sp. nov. from the Albian Mzinene Formation, KwaZulu-Natal, South Africa. *Afr. Nat. Hist.* **5**: 43–47.

—— & —— 2010. Cretaceous faunas from Zululand and Natal, South Africa. The ammonite subfamily Acanthoceratinae de Grossouvre, 1894. *Afr. Nat. Hist.* **6**: 1–76.

—— & —— 2011a. Cretaceous faunas from Zululand and Natal,South Africa. The ammonite genus *Oxytropidoceras* Stieler, 1920. *Afr. nat. Hist.* **7**: 69–102.

—— & —— 2011b. Cretaceous faunas from Zululand and Natal, South Africa. The ammonite subgenus *Hauericeras (Gardeniceras)* Matsumoto & Obata, 1955. *Palaeont. afr.* **12**: 43–58.

—— & —— 2012a. Cretaceous faunas from Zululand and Natal, South Africa. The ammonite genus *Codazziceras* Etayo-Serna, 1979. *Palaeont. afr.* **12**: 1–2.

—— & —— 2012b. The ammonite genus *Diaziceras* Spath, 1921, from the Campanian of KwaZulu-Natal, South Africa, and Madagascar. *Palaeont. afr.* **12**: 3–14.

—— & —— 2012c. A new species of the ammonite genus *Salaziceras* Breistroffer,1936, from the Lower Cenomanian Mzinene Formation. *Palaeont. afr.* **12**: 15–18.

—— & —— 2012d. Cretaceous faunas from Zululand and Natal, South Africa.The ammonite genera *Mojsisovicsia* Steinmann, 1881, *Dipoloceroides* Breistroffer, 1947, and *Falloticeras* Parona & Bonarelli, 1897. *Afr. nat. Hist.* **8**: 1–15.

—— & —— 2012e. Cretaceous faunas from Zululand and Natal, South Africa.The desmoceratoid ammonite genera *Moretella* Collignon, 1963, *Beudanticeras* Hitzel, 1902, and *Aioloceras* Whitehouse, 1926. *Afr. nat. Hist.* **8**: 55–75.

—— & —— 2013a. Cretaceous faunas from Zululand and Natal, South Africa. The ammonite subfamily Stoliczkaiinae Breistroffer, 1953. *Afr. nat. Hist.* **8**: 55–75.

—— & —— 2013b. Scaphitid ammonites from the Upper Cretaceous of KwaZulu-Natal and Eastern Cape Province, South Africa. *Acta Geol. Polonica* **63**(4): 527–543.

—— & —— 2013c. Cretaceous faunas from Zululand and Natal, South Africa. New records of Maastrichtian ammonites of the Family Kossmaticeratidae. *Afr. nat. Hist.* **9**: 55–59.

—— & —— 2013c. Cretaceous faunas from Zululand and Natal, South Africa. *Texasia cricki* Spath, 1921 (Cephalopoda: Ammonoidea) an early Santonian marker fossil from the Mzamba Formation of the Eastern Cape Province. *Palaeont. afr.* **34**: 55–59.

—— & —— 2014a. Cretaceous faunas from Zululand and Natal,South Africa. *Valdedorsella*, *Pseudohaploceras*, *Puzosia*, *Bhimaites*, *Pachydesmoceras*, *Parapuzosia (Austiniceras)* and *P. (Parapuzosia)* of the ammonite subfamily Puzosiinae Spath, 1922. *Afr. nat. Hist.* **10**: 1–46.

—— & —— 2014b. Cretaceous faunas from Zululand and Natal,South Africa. The ammonite subfamily Mantelliceratinae Hyatt, 1903. *Afr. nat. Hist.* **11**: 1–42.

—— & —— 2014c. Cretaceous faunas from Zululand and Natal, South Africa. The Albian ammonite genus *Douvilleiceras* de Grossouvre, 1894. *Afr. nat. Hist.* **11**: 42–82.

——, —— & Mateer, N. 1987. First record of an Upper Cretaceous sauropod dinosaur from Zululand, South Africa. *S. Afr. J. Sci.* **83**: 173–174.

——, Walaszczyk, I. & Klinger, H.C. 2008. *Cladoceramus* (Bivalvia, Inoceramidae) - ammonite associations from the Santonian of KwaZulu, South Africa. *Cret. Res.* **28**: 267–293.

——, Wright, C. W. & Klinger, H. C. 1979. Cretaceous faunas from Zululand and Natal, South Africa. A new genus and species of tuberculate desmoceratacean ammonite from the Mzinene Formation (Albian). *Ann. S. Afr. Mus.* **78**: 29–38.

——, —— & —— 1983.Cretaceous faunas from Zululand and Natal, South Africa:

The ammonite subfamily Barroisiceratinae Basse, 1947. *Ann. S. Afr. Mus.* **90**: 241–324.

KENT, P. E., HUNT, J. A., JOHNSTONE, D. W., 1971. The geology and geophysics of coastal Tanzania. *Inst. Geol. Sc., Geophys. Pap.* **6**: i–vi, 1–101.

KIEL, S. & BANDEL, K. 1999. The Pugnellidae, a new stromboidean family (Gastropoda) from the Upper Cretaceous. *Paläont. Zt.* **73**(1/2): 47–58.

KILIAN, W. 1902a. Sur la presence de l'etage Aptien dans le sudest de l'Afrique. *C. r. hebd. Seanc. Acad. Sci., Paris* **75**: 68.

—— 1902b. Ueber Aptien in Sudafrika. *Zentr. Miner. Geol. Paläont.* **1902**: 465–468.

KING, L. C. & MAUD, R. M. 1964. The geology of Durban and environs. *Bull. geol. Surv. S. Afr.* **42**: 1–49.

KIRCHHEIMER, F. 1934. Pollen from the Upper Cretaceous dysodil of Banke, Namaqualand (South Africa). *Trans. r. Soc. S. Afr.* **21**: 41–50.

KITCHIN, F. L. 1908. The invertebrate fauna and palaeontological relations of the Uitenhage Series. *Ann. S. Afr. Mus.* **7**: 21–250.

KLINGER, H. C. 1976. Cretaceous heteromorph ammonites from Zululand. *Mem. geol. Surv. S. Afr.* **69**:1–142.

—— 1989. The ammonite subfamily Labeceratinae Spath 1925: Systematics, phylogeny, dimorphism and distribution (with a description of a new species). *Ann. S. Afr. Mus.* **98**: 189–219.

—— & KENNEDY, W. J. 1977. Cretaceous faunas from Zululand, South Africa, and southern Mozambique: The Aptian Ancyloceratidae (Ammonoidea). *Ann. S. Afr. Mus.* **73**: 215–359.

—— & —— 1978. Turrilitidae (Cretaceous Ammonoidea) from South Africa, with a discussion of the evolution and limits of the family. *J. mollusc. Stud.* **44**: 1–48.

—— & —— 1979. Cretaceous faunas from Southern Africa. Lower Cretaceous ammonites, including a new bochianitid genus, from Umgazana, Transkei. *Ann. S. Afr. Mus.* **78**: 11–19.

—— & —— 1980a. Cretaceous faunas from Zululand and Natal, South Africa: The ammonite subfamily Texanitinae Collignon, 1948. *Ann. S. Afr. Mus.* **80**: 1–357.

—— & —— 1980b. Cretaceous faunas from Zululand and Natal, South Africa: A new sex-tuberculate texanitid. *Ann. S. Afr. Mus.* **82**: 321–331.

—— & —— 1980c. The Umzamba Formation at its type section Umzamba estuary (Pondoland, Transkei), the ammonite content and palaeogeographical distribution. *Ann. S. Afr. Mus.* **81**: 207–222.

—— & ——, LEES, J.A. & KITTO, S. 2001. Upper Maastrichtian ammonites and nannofossils, and a Palaeocene nautiloid from Richards Bay, KwaZulu, SouthAfrica. *Acta Geol. Polonica* **51**: 273–291.

—— & —— 1989. Cretaceous faunas from Zululand and Natal, South Africa: The ammonite family Placenticeratidae Hyatt, 1900; with comments on the systematic position of the genus *Hypengonoceras* Spath, 1924. *Ann. S. Afr. Mus.* **98**: 241–408.

—— & —— 1993. Cretaceous faunas from Zululand and Natal, South Africa. The ammonite genus *Eubaculites* Spath, 1926. *Ann. S. Afr. Mus.* **102**(6): 185–264.

—— & —— 1997. Cretaceous faunas from Zululand and Natal, South Africa. The ammonite family Baculitidae Gill, 1871 (excluding the genus *Eubaculites*). *Ann. S. Afr. Mus.* **105**(1): 1–206.

—— & —— 2001. Stratigraphic and geographic distribution, phylogenetic trends and general comments on the ammonite family Baculitidae Gill, 1871 (with an annotated list of species referred to the family). *Ann. S. Afr. Mus.* **107**(1): 1–290.

—— & —— 2008. *Mkuzeiella andersoni* gen. et sp. nov. (Cephalopoda, Ammonoidea) from the Albian Mzinene Formation of KwaZulu-Natal, South Africa. *Bull. Inst. r. Sci. Nat. Belg. (Sciences de la Terre)* **78**: 179–191.

——, —— & DINGLE, R. V. 1972. A Jurassic ammonite from South Africa. *N. Jb. Geol. Paläont., Mh.* **11**: 653–659.

——, —— & SIESSER, W. G. 1976. *Yabeiceras* (Coniacian ammonite) from the Alphard Group off the southern Cape coast. *Ann. S. Afr. Mus.* **69**: 161–168.

——, WIEDMANN, J. & KENNEDY, W. J. 1975. A new carinate phylloceratid ammonite from the early Albian (Cretaceous) of Zululand, South Africa. *Palaeontology* **18**: 657–664.

——, ——, —— 1983. Cretaceous faunas from Zululand and Natal, South Africa: The ammonite subfamily Barroisiceratinae Basse, 1947. *Ann. S. Afr. Mus.* **90**: 241–324.

—— & LOCK, B. E. 1979. Upper Cretaceous sediments from the Igoda River mouth, East London, South Africa. *Ann. S. Afr. Mus.* **77**: 71–83.

—— & MALCHUS, B. E. 2005. The first record of *Agerostrea ungulata* (von Schlotheim, 1813) (Bivalvia: Ostreoidea) from the Upper Maastrichtian of KwaZulu, South Africa, with a discussion of its distribution in southeast Africa and Madagascar. *Afr. nat. Hist.* **4**: 11–16.

KOLLMAN, H. A. 2009. A Late Cretaceous Aporrhaidae-dominated gastropod assemblage from the Gosau Group of the Pletzach Alm near Kramsach (Tyrol, Austria). With an appendix on the taxonomy of Mesozoic Aporrhaidae and their position in the superfamily Stromboidea. *Ann. Naturhist. Mus. Wien* **111A**: 33–72.

KOUNOV, A., VIOLA, G., DE WIT, M. J. & ANDREOLI, M. 2013. A Mid Cretaceous paleo-Karoo river valley across the Knersvlakte plain (north-western coast of South Africa): evidence from apatite fission-track analysis. *S. Afr. J. Geol.* **111**: 409–420.

KRAUSE, D. W., GOTTFRIED, M. D., O'CONNOR, P. M. & ROBERTS, E. M. 2003. A Cretaceous mammal from Tanzania. *Acta palaeont. Polonica* **48**(3): 321–330.

KRAUSS, F. 1843. Uber die geologischen Verhaltnisse der ostlichen Kuste des Kaplandes. *Amtl. Ber. Gesell. Dtsch. Naturfor. Aetze* **1843**: 126–132.

—— 1850. Ueber einige Petrefakten aus der untern Kreide des Kaplandes. *Nova Acta Acad. Caesar. Leop.-Carol. Nat. Cur.* **22**(2): 439–464.

KRENKEL, E. 1910a. Die Aptfossilien der Delagoa-Bai (Sudostafrika). *Neues Jb. Miner. Geol. Paläont.* **1**: 142–168.

—— 1910b. Die untere Kreide von Deutsch-Östafrika. *Beitr. Paläont. Geol. Ost -Ung.* **33**: 201.

LACEY, W. S. 1961. Report on fossils from Chalala and Mangulane in the province of Moçambique. *Bolm Servs geol. Min. Lourenço Marques* **27**: 7–15.

LAMBERT, G. 1972. *A study of Cretaceous foraminifera from northern Zululand, South Africa.* Unpubl. M.Sc. thesis, Univ. Natal, Durban.

—— & SCHEIBNEROVA, V. 1974. Albian Foraminifera of Zululand (South Africa) and Great Artesian Basin (Australia). *Micropalaeont.* **20**: 76–96.

LANG, W. D. 1908. Polyzoa and Anthozoa from the Upper Cretaceous limestone of Need's Camp, Buffalo River. *Ann. S. Afr. Mus.* **7**: 1–11.

LENZ, O. 1877. Petrefakten von der Loango-Küste (west-Afrika). *Verh. k. k. Geol. Reichsanst.* **6**: 278–279.

—— 1878. Geologische Mittheilungen aus Westafrika. *Verh. k. k. geol. Reichsanst.* **7**: 148–153.

LITTLE, J. DE V. 1957. A new species of *Trigonia* from Upper Cretaceous beds near the Itongazi River, Natal. *Palaeont. afr.* **4**: 117–121.

LORIOL, P. DE. 1888. Notes sur la géologie de la province d'Angola. *Archs Sci. phy. Nat.* **19**: 67–71.

MADEL, E. 1960. Monimiacean-Holzer aus der oberkretazischen Umzambaschichten von Ost-Pondoland, S. Afrika. *Senckenberg. lethaia* **41**: 331–391.

MAKRIDES, M. 1979. *Micropalaeontology of the Upper Cretaceous Mzamba Formation, Pondoland, Transkei.* Unpubl. M.Sc. thesis, Univ. Witwatersrand.

MARTIN, H. 1973. Palaeozoic, Mesozoic and Cenozoic deposits on the coast of South-West Africa. *In*, BLANT, G., ed.. *Sedimentary basins of the African coasts. 2nd Part, South and East Coasts*: 7–15. Assoc. Afr. geol. Survs: Paris.

MASSEA, P. & LAURENT, O. 2016. Geological exploration of Angola from Sumbe to Namibe: a review at the frontier between geology, natural resources and the history of geology. *C. R. Geoscience* **348**: 80–88.

MATEER, N. J. 1987. A new report of a theropod dinosaur from South Africa. *Palaeontology* **30**: 141–145.

MAUFE, H.B. 1922. The Karroo and Kalahari Beds of the Gwampa Valley, Bubye district. *Trans. Rhod. Sci. Assoc.* **20**: 1–10.

MATEUS, O., MORAIS, M.L., SCHULP, A.S., JACOBS, L.L. & POLCYN, M.J. 2006. The Cretaceous of Angola. *J. vert. Paleont.* **26**(3): 96A–97A.

——, JACOBS, L.L., POLCYN, M.J., SCHULP, A.S., NETO, A.B. AND ANTUNES, M.T. 2008: Dinosaur and turtles from the Turonian of Iembe, Angola. *Livr. Res. "Tercer Congreso Latinoamericano de Paleontología de Vertebrados", Neuquén, Argentina*: 156.

——, ——, ——, ——, VINEYARD, D.P., NETO, A.B. & ANTUNES, M.T. 2009. The oldest African eucryptodiran turtle from the Cretaceous of Angola. *Acta palaeont. Polon.* **54**(4): 581–588.

——, ——, SCHULP, A. S., POLCYN, M. J., TAVARES, T. S., NETO, A. B., MORAIS, M. L. & ANTUNES, M. T. 2011. *Angolatitan adamastor*, a new sauropod dinosaur and the first record from Angola. *Anais Acad. Brasil. Ciênc.* **83**(1): 1–13.

——, POLCYN, M. J., JACOBS, L. L., ARAUJO, J., SCHULP, A. S., MARINHEIRO, J., PEREIRA, B. & VINEYARD, D. 2012. Cretaceous amniotes from Angola: dinosaurs, pterosaurs, mosasaurs, plesiosaurs and turtles. *Actas V J. Int. Paleont. Dinosaur. Ent., Sal Inf. Burgos*: 71–105.

MCGOWRAN, B. & MOORE, A. C. 1971. A reptilian tooth and Upper Cretaceous microfossils from the Lower Quarry at Need's Camp, South Africa.*Trans. geol. Soc. S. Afr.* **72**(4): 103–105.

MCLACHLAN, I. R., BRENNER, F. W. & MCMILLAN, I. K. 1976. The stratigraphy and micro-palaeontology of the Cretaceous Brenton Formation and the PB-A/1 well, near Knysna, Cape Province. *Trans. geol. Soc. S. Afr.* **79**(3): 341–370.

—— & MCMILLAN, I. K. 1976. Review and stratigraphic significance of Southern Cape Mesozoic palaeontology. *Trans. geol. Soc. S. Afr.* **79**:197–212.

—— & —— 1979. Microfaunal biostratigraphy, chronostratigraphy and history of Mesozoic and Cainozoic deposits on the coastal margin of South Africa. *Spec. Publ. geol. Soc. S. Afr.* **6**: 161–181.

MCMILLAN, I. K. 1999. The Foraminifera of the Late Valanginian to Hauterivian (Early Cretaceous) Sundays River Formation of the Algoa Basin, Eastern Cape Province, South Africa. *Ann. S. Afr. Mus.* **106**: 1–120.

MEISTER, C., BUTA, A., DAVID, B. & TAVARES, T. 2011. Les ammonites de la limite Albo-Cénomanien dans la région de Sumbe (bassin de la Kwanza, Angola). *Rev. Paleobiol.* **30**(2): 685–781.

MEUNIER, S. 1887. Contribution à la géologie de l'Afrique occidentale. *Bull. Soc. géol. Fr.* **16**: 61–68.

MILLER, A. K. & CARPENTER, L. B. 1956. Cretaceous and Tertiary nautiloids from Angola. *Estud. Ens. Docum. Jta Invest. Ultramar* **21**: 1–48.

MONTANARO-GALLITELLI, E. & LANG, Z. 1937. Celenterati, Echinodermi e Brachiopodi del Cretacico medio-superior dello Zululand. *Palaeontogr. ital.*, n.s., **37**: 193–310.

MOUTA, F. & BORGES, A. 1926. Sur le Crétacé du littorale de l'Angola (districts de Benguela et Mossamedes). *Bolm Ag. ger. Colón. Ultramar* **14**: 30–55.

MUIR-WOOD, H. M. 1953. Description of a new species of *'Terebratula'* from the Cretaceous of Zululand. *Trans. Proc. geol. Soc. S. Afr.* **56**: 183–186.

MÜLLER, G. 1900. Versteinerungen des Jura und der Kreide. *Deutsch-Ost-Afrika* **7**: 514–571.

MULLER-STOLL, W. R. & MADEL, E. 1972. Fossil woods of Monimiacea and Euphorbiacea from the Upper Cretaceous Umzamba Beds of east Pondoland. *Trans. geol. Soc. S. Afr.* **65**: 93–104.

NETO, M. G. M. 1960a. Novidades paleontológicas. *Bolm Servs geol. Min. Angola* **2**: 73

—— 1960b. Géologie de la région - Benguela-Cuio. *Bolm Serv. geol. min. Angola* **1**: 89–99.

—— 1961. As baçias sedimentaires de Benguele e Moçâmedes. *Bolm Serv. geol. min. Angola* **3**: 63–93.

NEUMAYR, M. 1885. *Phylloceras semistriatum* Orb. von Mossambique. *Denkschr. Akad. Wiss. Wien* **50**: 139.

NEWTON, R. B.1896. On the occurrence of *Alectryonia ungulata* in SE Africa; with a notice of the previous researches on the Cretaceous conchology of Southern Africa. *J. Conchol.* **8**: 136–151.

—— 1909. Cretaceous Gastropoda and Pelycypoda from Zululand. *Trans. r. Soc. S. Afr.* **1**: 1–106.

—— 1916. On some Cretaceous Brachiopoda and Mollusca from Angola, Portuguese West Africa. *Trans. roy. Soc. Edinb.* **51**: 561–580.

—— 1920. On some African freshwater fossils from central South Africa. *Ann. Mag. nat. Hist.* **5**(9): 241–249.

—— 1924. A contribution to the palaeontology of Portuguese East Africa. *Trans. Proc. geol. Soc. S. Afr.* **26**: 141–159.

OVECHKINA, M. N., WATKEYS, M. & MOSTOVSKI, M. B. 2009. Calcareous nannofossils from the stratotype section of the Upper Cretaceous Mzamba Formation, Eastern Cape, South Africa. *Palaeont. afr.* **44**: 129–133.

PEARSON, P. N., NICHOLAS C.J., SINGANO, J. M., BOWN, P.R., COXALL, H. K., DONGEN, B. E. VAN, HUBERG,B. T., KAREGA, A., LEES, J. A., MSAKY, E., PANCOST, R. D., PEARSON, M., ROBERTS, A. P. 2004. Paleogene and Cretaceous sediment cores from the Kilwa and Lindi areas of coastal Tanzania: Tanzania Drilling Project Sites 1–5. *J. Afr. Earth Sci.* **39**: 25–62.

PHILLIPS, J. F. V. 1927. Fossil *Widdringtonia* in lignite of the Knysna Series with a note on fossil leaves of several other species. *S. Afr. J. Sci.* **24**: 188–197.

PIENAAR, R. N. 1969. Upper Cretaceous calcareous nannoplankton from Zululand, South Africa. *Palaeont. afr.* **12**: 75–128.

PLOWS, W. J. 1921. The Cretaceous rocks of Pondoland. *Ann. Durban Mus.* **3**: 58–66.

POLCYN, M. J., JACOBS, L. L., SCHULP, A. & MATEUS, O. 2007a. *Halisaurus* (Squamata: Mosasauridae) from the Maastrichtian of Angola. *J. vert. Paleont.* **27**(3): 130A.

——, ——, —— & —— 2007b. Morphology and systematic position of *Angolasaurus bocagei* and the evolution of the braincase in Plioplatecarpine mosasaurs. *Publ. Abstr. 2nd Mosasaur Meet. Sternberg Mus.,Hays, Kansas*: 20.

——, ——, —— & —— 2007c. The mosasaurs of Angola. *Publ. Abstr. 2nd Mosasaur Meeting, Sternberg Mus., Hays, Kansas*: 21.

——, ——, —— & —— 2010. The North African mosasaur *Globidens phosphaticus* from the Maastrichtian of Angola. *Hist. Biol.* **22**(1–3): 175–185.

——, ——, MATEUS, O. & SCHULP, A. 2009. New specimens of *Angolasaurus bocagei* and comments on the early radiations of plioplatecarpine mosasaurs. *J. vert. Paleont.* **29**(3): 165A.

QUENNEL, A. M., MCKINLEY, A. C. M. & AITKIN, W. G. 1956. Summary of the geology of Tanganyika. I. Introduction and stratigraphy. *Mem. geol. Surv. Tanganyika* **1**(1): 1–264.

REAL, F. 1966. Geologia da bacia do Rio Zambezi (Mocambicano) characteristicas geologico mineiras da bacia do rio Zambeze em territorio mocambicano. *Jta Invest. Ultramar* **1966**: 1–183.

RENNIE, J. V. L. 1929. Cretaceous fossils from Angola (Lamellibranchia and Gastropoda). *Ann. S. Afr. Mus.* **28**: 1–54.

—— 1930. New Lamellibranchia and Gastropoda from the Upper Cretaceous of Pondoland (with an appendix on some species from the Cretaceous of Zululand). *Ann. S. Afr. Mus.* **28**: 1–54.

—— 1931. Note on fossil leaves from the Banke Clay. *Trans. r. Soc. S. Afr.* **19**: 251–253.

—— 1934. Two new species of *Pleurotomaria* from the Lower Cretaceous Uitenhage Series. *S. Afr. J. Sci.* **31**: 233–235.

—— 1935a. Upper Cretaceous Lamellibranchia from Incomanini, Portuguese East Africa. *Ann. Trans. Mus.* **18**: 325–347.

—— 1935b. On a new species of *Lysis* (Gastropoda) from the Cretaceous of Pondoland. *Rec. Albany Mus.* **4**: 244–247.

—— 1936. Lower Cretaceous Lamellibranchia from northern Zululand. *Ann. S. Afr. Mus.* **31**: 277–391.

—— 1937. Fossils from the Lebombo volcanic formation. *Bolm. Serv. Indust. Serv. geol. Lourenço Marques* **1**: 1–25.

—— 1943. Fauna do Cretaçico Superior de Grudja. *Bolm. Servs. Indust. Min. geol. Lourenço Marques* **5**: 1–44.

—— 1944. Fossils from Mabosi Conglomerate of Lourenço Marques. *Bolm. Servs. Indust. Min. geol. Lourenço Marques* **6**: 27–35.

—— 1945. Lamelibrânquios Gastrópodos do Cretácico Superior de Angola. *Mem. Jta geogr. colon.* **1**: 1–141.

——1947. Aptian fossils from Chalala near Lourenço Marques. *Bolm. Servs. geol. Min. Moçambique.* **9**: 45–81.

Reuning, E. 1931. A contribution to the geology and palaeontology of the western edge of the Bushmanland Plateau. *Trans. r. Soc. S. Afr.* **19**: 215–232.

Reyment, R. A. & Tait, E. A. 1972. Biostratigraphical dating of the early history of the South Atlantic Ocean. *Phil. Trans. R. Soc., B*, 264–295.

Rich, T. H. V., Molnar, R. E. & Rich, P. V. 1983. Fossil vertebrates from the Late Jurassic or Early Cretaceous Kirkwood Formation, Algoa Basin, Southern Africa. *Trans. geol. Soc. S. Afr.* **86**: 281–291.

Rigassi, D. A. & Dixon, G. E. 1972. Cretaceous of the Cape Province, Republic of South Africa. *Proc. Ibadan Conf. Afr. Geol.* **1970**: 513–527.

Rogers, A. W. 1906. Geological survey of parts of the division of Uitenhage and Alexandria, with appendix on the occurrence of wood beds on Loerie and Gamtoos Rivers. *Ann. Rep. geol. Comm. Cape Good Hope* **1905**: 11–46.

—— 1909. Notes on a journey to Knysna. *13th Ann. Rep. geol. Comm. Cape Good Hope* **1908**: 129–134.

—— 1910. The Swartkops borehole. *Ann. Rep. geol. Comm. Cape Good Hope* **1909**: 111–116.

—— 1916. The geology of part of Namaqualand. *Trans.Proc. geol. Soc. S. Afr.* **18**: 72–101.

—— & Schwarz, E. H. L. 1902. Geological survey of the rocks in the southern parts of the Transkei and Pondoland, including a description of the Cretaceous rocks of E. Pondoland. *Ann. Rep. geol. Comm. Cape Good Hope* **1902** (for 1901): 38–46.

Roman, J. & Goncalves, E. 1965. Echinides du Crétacé et du Miocene de Moçambique. *Garçia de Orta* **13**(3): 267–278.

Romanes, M. F. Notes on an algal limestone from Angola. *Trans. roy. Soc. Edinb.* **51**: 581–584.

Ross, C., Sues, H.-D. & De Klerk, W. J. 1999. Lepidosaurian remains from the Lower Cretaceous Kirkwood Formation of South Africa. *J. Vert. Paleont.* **19**: 21–27.

Rust, I. C. & Winter, H. de la R. 1979. The Cretaceous Oudshoorn basin. *In*, Rust, I. C., ed., *Geokongr.79, Excursion guide book*: 53–61. E.H. Walton & Co.: Port Elizabeth.

SACS 1980. Stratigraphy of South Africa Part 1. Lithostratigraphy of the Republic of South Africa, South West Africa/Namibia and the Republic of Bophuthatswana, Transkei and Venda. *Handbk geol. Surv. S. Afr.* **8**: 1–690.

Schelpe, E. A. C. L. E. 1955. *Osmundites natalensis* – a new fossil fern from the

Cretaceous of Zululand. *Ann. Mag. nat. Hist.* **8**(12): 652–656.

SCHLÖSSER, M. 1928. Uber Tertiar und obere Kreide aus Portugiesisch-Ostafrika. *Abh. Bayer. Akad. Wiss. math. nat. wiss. Abt.* **32**(2): 1–25.

SCHOLZ, A. 1985. The palynology of the upper lacustrine sediments of the Arnot Pipe, Banke, Namaqualand. *Ann. S. Afr. Mus.* **95**(1): 1–109.

SCHULP, A. S., MATEUS, O., POLCYN, M. J. & JACOBS, L. L. 2006. A new *Prognathodon* (Squamata: Mosasauridae) from the Cretaceous of Angola. *J. vert. Paleont.* **26**(3): 122A.

——, ——, ——, —— & MORAIS M. L. 2008. A new species of *Prognathodon* (Squamata, Mosasauridae) from the Maastrichtian of Angola, and the affinities of the mosasaur genus *Liodon*. *In*, Everhart M.J., ed., *Proc. 2nd Mosasaur Meet. Fort Hays Stud., Spec. Issue* **3**: 1–12.

——, POLCYN, M. J., MATEUS, O., JACOBS, L. L., MORAIS M. L. & DA SILVA TAVARES, T. 2006. New mosasaur material from the Maastrichtian of Angola, with notes on the phylogeny, distribution and palaeoecology of the genus *Prognathodon*. *Publ. naturhist. Genoots. Limburg* **45**(1): 57–67.

SCHULTZE-MOTEL, J. 1966. Ergebnisse der Forschungsreise Richard Krausel's nach Sud und Sudwest Afrika. I. 9: Gymnospermen Holzer aus den Oberkretazischen Umzamba Schichten von Ost Pondoland (S. Afrika). *Senckenberg. leth.* **47**: 279–337.

SCHUTTE, I. C. 1974. 'n Nuwe voorkoms van die Formasie Malvernia suid van Pafuri, Nationale Krugerwildtuin. *Ann. geol. Surv. S. Afr.* **9**: 83–84.

SCHWARZ, E. H. L. 1900. Knysna, between the Gouwkanna (Homtini) and the Blue Krantz Rivers. *Ann. Rep. geol. Comm. Cape Good Hope* **1899**: 51–64.

—— 1915. New Cretaceous fossils from Brenton, Knysna. *Rec. Albany Mus.* **2**: 120–126.

SCOTT, L. 1971. Lower Cretaceous pollen and spores from the Algoa basin (South Africa). Unpubl. M.Sc. thesis, Univ. O.F.S., Bloemfontein.

—— 1976. Palynology of Lower Cretaceous deposits from the Algoa basin (Republic of South Africa). *Pollen spores* **18**: 563–609.

SEWARD, A.C. 1903. The fossil floras of Cape Colony. *Ann. S. Afr. Mus.* **4**: 1–122.

—— 1907. Notes on fossil plants from South Africa. *Geol. Mag.*, n.s., **4**(5): 481–487.

SHARPE, D. 1856. Description of fossils from the secondary rocks of Sundays River and Zwartkops River, South Africa, collected by Dr. Atherstone and A. G. Baine Esq. *Trans. geol. Soc. Lond.* **7**: 193–203.

SHONE, R. W. 1976. *The stratigraphy and sedimentology of the Sundays River Formation*. Unpubl. M. Sc. thesis, Univ. Port Elizabeth.

—— 1978. The case for lateral gradation between the Kirkwood and Sundays River Formations, Algoa basin. *Trans. geol. Soc. S. Afr* **81**: 319–326.

SIESSER, W. G. 1982. Cretaceous calcareous nannoplankton in South Africa. *J. Paleont.* **56**: 335–350.

SILVA, G. H. DA. 1961. Ammonite nouvelle du Campanien de la Barra do Dande (Angola). *Mem. Notic. Mus. miner. geol. Univ. Coimbra* **51**: 19–25.

SILVA, G. H. da. 1962. Ammonites do Cretacici Inferior do Rio Maputo (Catuane - Moçambique). *Bolm Servs geol. Min. Lourenço Marques* **29**: 7–32.

—— 1963a. O género *Anacorax* no Cretácico Superior de Angola. *Mem. Notic. Mus. miner. geol. Univ. Coimbra* **55**: 25–41.

—— 1963b. Sobre os lamelibrânquios do Cretácico da regiâo de Carunjamba – Salinas – S. Nicolau. *Mem. Notic. Mus. miner. geol. Univ. Coimbra* **56**: 27–34

—— 1965a. Contribuição para o estudo dos lamelibrânquios Cretácicos da região de Moçâmedes. *Bolm Servs geol. Min.* **11**: 137–168.

—— 1965b. Contribuição para von catalogo dos lamelibranquios fosseis do Cretácico de Moçambique. *Rev. estud. Ger. Univ. Moçambique* (VI) **2**(2): 1–40.

—— 1966. Contribuição para um catalogo dos gastropodes fosseis de Moçambique. *Rev. estud. Ger. Univ. Moçambique* (VI) **3**: 71–100.

—— 1972. Notas de Paleontologia Mocambicana. I. *Pinna robinaldina* d'Orbigny do Cretácico Inferior de Moçambique. *Rev. Cienc. Geol., Lourenço Marques* (A) **5**: 1–8.

SIEDNER, G. & MITCHELL, J. G. 1976. Episodic Mesozoic volcanism in Namibia and Brazil: a K-Ar isochron study bearing on the opening of the South Atlantic. *Earth Planet. Sci. Letters* **30**: 292–302.

SIEVERTS-DORECK, H. 1939. Jura- und Kreide-Crinoideen aus Deutsch-Ostafrika. *Palaeontographica*, Suppl. 7, II Reihe, Teil II, Lfg **3**: 217–231.

SILVA, G. H. DA 1966. Sobre a ocorrencia do Jurassico marinho no norte de Mozambique. *Rev. Estud. Ger. Univ. Mozambique*, **3**(2): 61–68.

SMITTER, Y. H. 1956. Foraminifera from the Upper Cretaceous beds occurring near the Itongazi River, Natal. *Palaeont. afr.* **3**: 103–107.

—— 1957.Upper Cretaceous Foraminifera from Sandy Point, St Lucia Bay, Zululand. *S. Afr. J. Sci.* **53**: 195–201.

SOARES, A. F. 1958. Sobre alguns fósseis da regiâo de entre Lobito e Catumbela (Angola). *Mem. Notic. Mus. miner. geol. Univ. Coimbra* **46**: 11–22.

—— 1959. Contribuição para o estudo da fauna fóssil da regiâo de entre Lobito e Catumbela (Angola). *Garçia de Orta* **7**: 135–154.

—— 1961. Lamelibrânquios do Cretácico da regiâo de Benguela-Cuio. *Bolm Servs. geol. min. Angola Publ. Mus. Lab. Min. Geol., Univ. Coimbra, Mem. e Not.* **55**: 1–24.

—— 1963a. Paleontologia de Angola. I. Sobre os lamelibrânquios cretácicos da regiâo de Benguela-Cuio. *Mem. Notic. Mus. miner. geol. Univ. Coimbra* **55**: 1–22.

—— 1963b. Alguns lamelibrânquios cretácicos da regiâo de entre o Posto de S. Nicolau e a Mulola du Caniço a sul do Chapéu Armado. *Mem. Notic. Mus. miner. geol. Univ. Coimbra* **56**: 35–40.

—— 1965. Contribuição para o estudo dos lamelibrânquios Cretácicos da região de Moçâmedes. *Bolm Servs geol. Min.* **11**: 137–168.

—— & SILVA, G. H. DA. 1969. Sobre a ocorrencia do Senoniano da margem esquerda do Rio Maputo (regiâo de Madubula). *Rev. Cienc. Geol. Lourenço Marques* (A) **2**: 1–7.

—— & —— 1970. Contribuição para o estudo da geologia do Maputo. Estratigrafia e paleontologia da regiâo de Madubula e suas relacoes com areas vizinhas. *Rev. Cienc. Geol. Lourenço Marques* (A) **3**: 1–85.

SOCIN, C. 1939. Gasteropodi e Lammelibranchi del Cretaceo medio-superiore dello Zululand. *Palaeontogr. ital.* **40**: 21–38.

SOHL, N. F. & KOLLMAN, H. A. 1985. Cretaceous actaeonellid gastropods from the Western Hemisphere. *Prof. Pap. U. S. geol. Surv.* **1304**: 1–104.

SORNAY, J. 1951. Ammonites albiennes et sénoniennes de l'Angola et de l'Afrique équatoriale française. *Rev. Zool. Bot. afr.* **44**: 271–277.

—— 1953. Ammonites nouvelles de l'Albien de Angola. *Rev. Zool. Bot. afr.* **47**: 52–59.

SPATH, L. F. 1921a. On Cretaceous Cephalopoda from Zululand. *Ann. S. Afr. Mus.* **12**: 217–321.

—— 1921b. On Upper Cretaceous Ammonoidea from Pondoland. *Ann. Durban Mus.* **3**: 39–57.

—— 1922a. On the Senonian ammonite fauna of Pondoland. *Trans. r. Soc. S. Afr.* **10**: 113–147.

—— 1922b. On Cretaceous Ammonoidea from Angola collected by Professor J. W. Gregory, D. Sc., F. R. S. *Trans. r. Soc. Edinb.* **53**: 91–160.

—— 1925. On Upper Albian Ammonoidea from Portuguese East Africa. *Ann.Transv. Mus.* **11**: 179–216.

—— 1930. On the Cephalopoda of the Uitenhage Beds. *Ann. S. Afr. Mus.* **28**: 131–157.

—— 1932. Review of 'H. Douvillé, 1931, Les ammonites de Salinas'. *Geol. Zetbl.* **1**: 24.

—— 1934. Review of 'Thiele, S., 1933, Neue Fossilfunde aus der Kreide von Angola mit einen Beitrag zur Stammesgeschichte der gattung *Pervinquieria* Böhm'. *Geol. Zentbl.* **32**: 201.

—— 1951. Preliminary notice on some Upper Cretaceous ammonite faunas from Angola. *Communções Servs geol. Port.* **32**(1): 123–130.

SPENCE, J. 1954. The geology of the Gabala coal field, Mbeya district, Tanganyika. *Bull. geol. Surv. Tanganyika* **25**: 1–34.

STAGMAN, J. 1978. The geology of Rhodesia. *Bull. geol. Surv. Rhod.* **80**: 1–126.

STAPLETON, R. P. 1975. Planktonic foraminifera and calcareous nannofossils at the Cretaceous-Tertiary contact in Zululand. *Palaeont. afr.* **18**: 53–70.

—— & BEER, E. M. 1977. Micropalaeontological age determination for the Brenton Beds. *Bull. geol. Surv. S. Afr.* **60**: 1–10.

SZAJNOCHA, L. 1884. Zur kenntnis der mittel cretacischen cephalopoden-fauna der Inseln Elobi an der Westküste Afrikas. *Denkschr. Akad. Wiss. Wien.* **49**: 231–238.

TATE, R. 1867. On some secondary fossils from South Africa. *Q. Jl geol. Soc. Lond.* **23**: 139–174.

TAVARES, T. 2005. *Ammonites et echinides de L'Albien du bassin de Benguela (Angola).* Ph.D thesis, Universite de Bourgogne.

—— MEISTER, C., DUARTE-MORAIS, M. L. DAVID, B. 2007. Albian ammonites of the Benguela Basin (Angola): a biostratigraphic framework. *S. Afr. J. Geol.* **110**: 137–156.

TEALE, E. O. 1924. The geology of Portuguese East Africa between the Zambesi and Sabi Rivers. *Trans. Proc. geol. Soc. S. Afr.* **26**: 103–129.

THIELE, S. 1933. Neue Fossilfunde aus der Kreide von Angola mit einen Beitrag zur Stammesgeschichte der Gattung *Pervinquieia Böhm. Zbl. Miner. geol. Paläont.* (B) **1933**: 110–123.

THOMPSON, R.W. 1977. Mesozoic sedimentation on the eastern Falkland Plateau. *In*, BARKER, P. *et al.*, eds, *Initial Rep. Deep Sea Drilling Project* **36**: 877–891.

VAIL, P. R., MITCHUM, R. M. & THOMPSON, S. 1977. Global cycles of relative changes of sea level. *Mem. Amer. Assoc. petrol. Geol.* **26**: 83–97.

VENTER, J. J. 1972. Type stratigraphy of the Sundays River Formation. *Unpubl. SOEKOR Rept.*

VENZO, S. 1936. Cefalopodi del Cretacea medio-superiore dello Zululand. *Palaeontogr. ital.* **36**: 59–133.

VINEYARD, D. P., JACOBS, L. L., POLCYN, M. J., MATEUS, O., SCHULP, A. S. & STRGANAC, C. 2009. *Euclastes* from the Maastrichtian of Angola and the distribution of the Angolachelonia. *Eugene Gaffney Turtle Symp.,Royal Tyrrell Mus.*

WACHENDORF, H. 1967. Unterkreide-Stratigraphie von Süd-Mozambique. *N. Jb. Geol. Paläont., Abh.* **129**: 272–303.

WALASZCZYK, I., KENNEDY, W., J. & KLINGER, H.C. 2009. Cretaceous faunas from Zululand and Natal, South Africa. Systematic palaeontology and stratigraphic potential of the Upper Campanian–Maastrichtian Inoceramidae (Bivalvia). *Afr. Nat. Hist.* **5**: 40–132

WAGNER, P. A. 1913. A traverse through the northern portion of the Mazoe district of Southern Rhodesia, into Portuguese territory. *Trans. Proc. geol. Soc. S. Af.* **15**: 124–139.

WEISSERMEL, W. 1900. Mesozoische und Känozoische Korallen aus Deutsch Ostafrika. *Deutsch-OstAfrika* **7**: 1–18.

WILLEY, L. E. 1973. Belennites from south-eastern Alexander Island: II. The occurrence of the family Belemnopseidae in the Upper Jurassic and Lower Cretaceous. *Bull. Brit. Antarctic Surv.* **36**: 33–59.

WINTER, H. DE LA R. 1973. Geology of the Algoa basin, South Africa. *In*, BLANT, G., ed.. *Sedimentary basins of the African coasts. 2nd Part, South and East Coasts*: 17–48. Assoc. Afr. geol. Survs: Paris.

—— 1979. Application of basic principles of stratigraphy to analysis of the Jurassic-Cretaceous interval in Southern Africa. *Spec. Publ. geol. Soc. S. Afr.* **6**: 183–196.

WOODS, H.1906. The Cretaceous fauna of Pondoland. *Ann. S. Afr. Mus.* **4**: 275–350.

—— 1908. Echinoidea, Brachiopoda and Lamellibranchia from the Upper Cretaceous limestones of Needs Camp, Buffalo River. *Ann. S. Afr. Mus.* **7**: 13–19.

WOODWARD, A. S. 1907. Notes on some Cretaceous fish teeth from the mouth of the Umpenyati River, Natal. *Rep. geol. Surv. Natal Zululand* **3**: 99–101.

ZWIEZYCKI, J. 1913. Zur Frage der unteren Kreide im Portugiesisch-Mozambique. *Sitz.-Ber. Ges. natur.-forsch. Freunde Berlin* **7**: 319–325.

—— 1914. Die Cephalopodenfauna der Tendaguruschichten in Deutsch-Ostafrika. *Arch. Biontol.* **4**(3): 7–97.

Illustrations

of

Cretaceous fossils

Plate 1

Berriasian-MiddleValanginian

Cycads, conifers, ferns, bivalve and belemnite
from the Kirkwood Formation

A. *Palaeozamia africana* (Tate).
B. *Araucarites rogersi* Seward.
C. *Conites* sp., x2.5.
D. *Otozamites rectus* (Tate)
E. *Cycadolepis jenkinsiana* (Tate).
F. *Sphenopteris fittoni* Seward, pinnule x3.
G. *Taeniopteris* sp.
H. *Cladophlebis denticulata atherstoni* Seward.
I. *Onychiopsis mantelli* (Brongniart), x1.5.
J. *Palaeozamia rubidgei* (Tate).
K. *Nilssonia tatei* Seward.
L. *Brachyphyllym* sp., a coniferous twig.
M. *Podozamites morrisii* (Tate).
N. *Cladophlebis browniana* (Dunker); pinnule x3.
O. *"Unio" uitenhagensis* Kitchin.
P. *Belemnopsis gladiator* Wyllie.

All x1 unless stated otherwise.

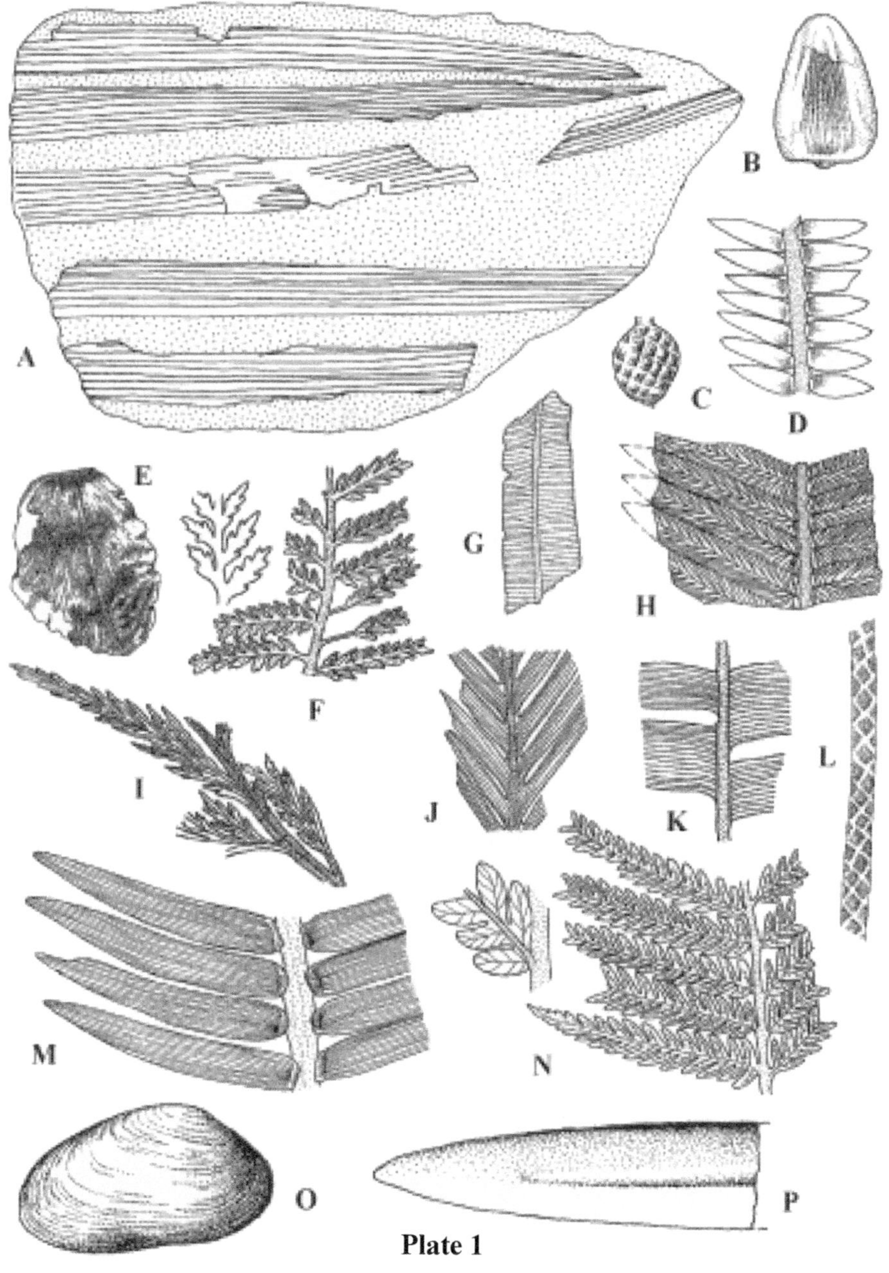

Plate 1

Plate 2

Berriasian-Aptian

Dinosaurs from the Kalahari Group, Kirkwood Formation and Karonga Formation

A. *Kangnasaurus coertzeei* Haughton, x0.25; a femur (after Cooper).
B. *Malawisaurus dixeyi* (Haughton), x0.25 (after Gomani).
C. *Nqwebasaurus thwazi* DeKlerk *et al.*, x2.5; a hindlimb.

All x1 unless stated otherwise.

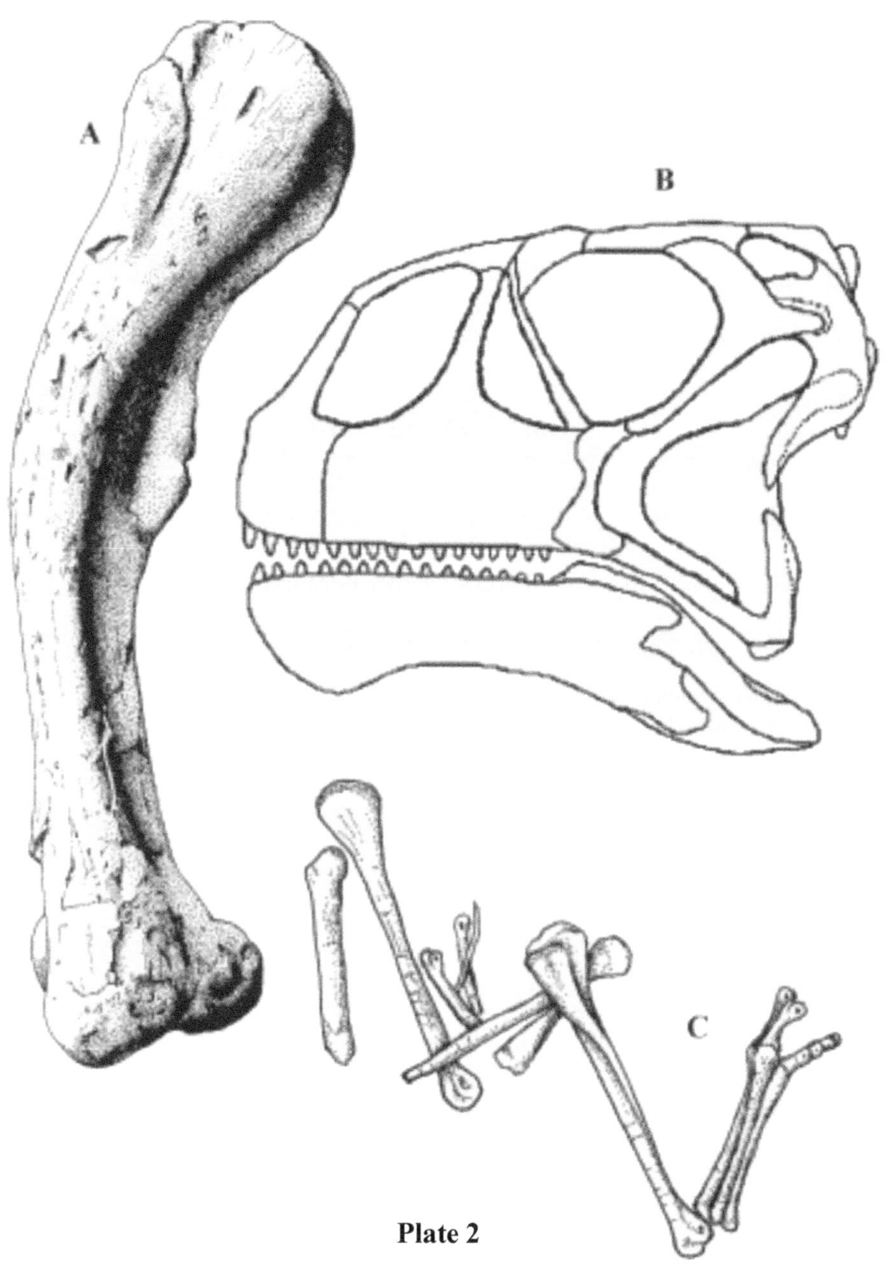

Plate 2

Plate 3

Late Valanginian

Bivalves, gastropods and shrimp
from the Sundays River Formation

A. *Steinmanella holubi* (Kitchin).
B. *Limnaea remota* Kitchin.
C. *Natica? mirifica* Kitchin, x2.
D. *Meyeria schwarzi* Kitchin.
E. *Aetostreon imbricatum* (Krauss).
F. *Monodonta hausmanni* Neumayr.
G. *Iotrigonia stowi* (Kitchin).
H. *"Turbo" rogersi* Kitchin, x2.

All x1 unless stated otherwise.

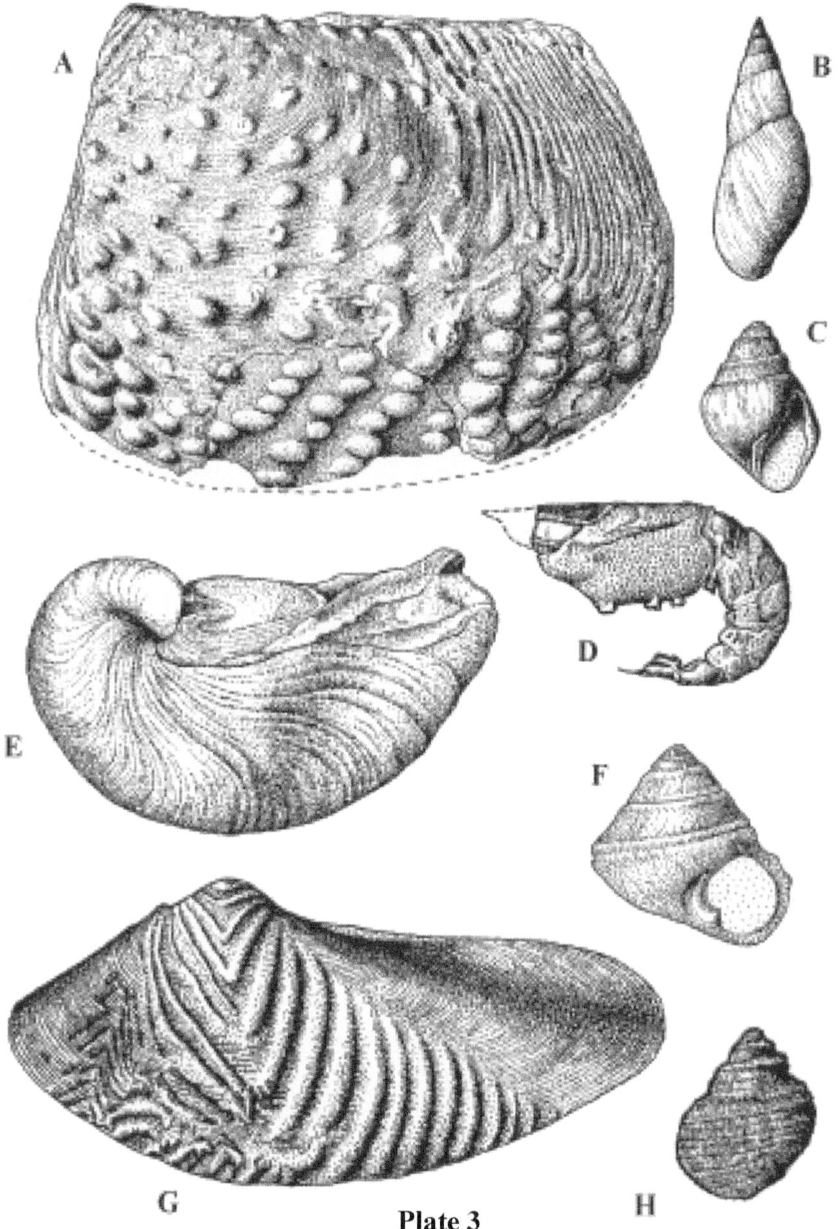

Plate 3

Plate 4

Late Valanginian

Bivalves and gastropods from the Sundays River Formation

A. *Pisotrigonia kraussii* (Kitchin).
B. *Inoperna baini* (Sharpe).
C. *Lycettia uitenhagensis* (Kitchin).
D. *Neritopsis? turbinata* Sharpe.
E. *Tancredia schwarzi* Kitchin.
F. *Chlamys?* cf. *subacutus* (Lamarck).
G. *Megacucullaea kraussi* (Tate).
H. *Trigonia tatei* Neumayr.
I. *Indogrammatodon jonesi* (Tate), x2.
J. *Arctica rugulosa* (Sharpe).
K. *Myopholas? dominicalis* (Sharpe).

All x1.

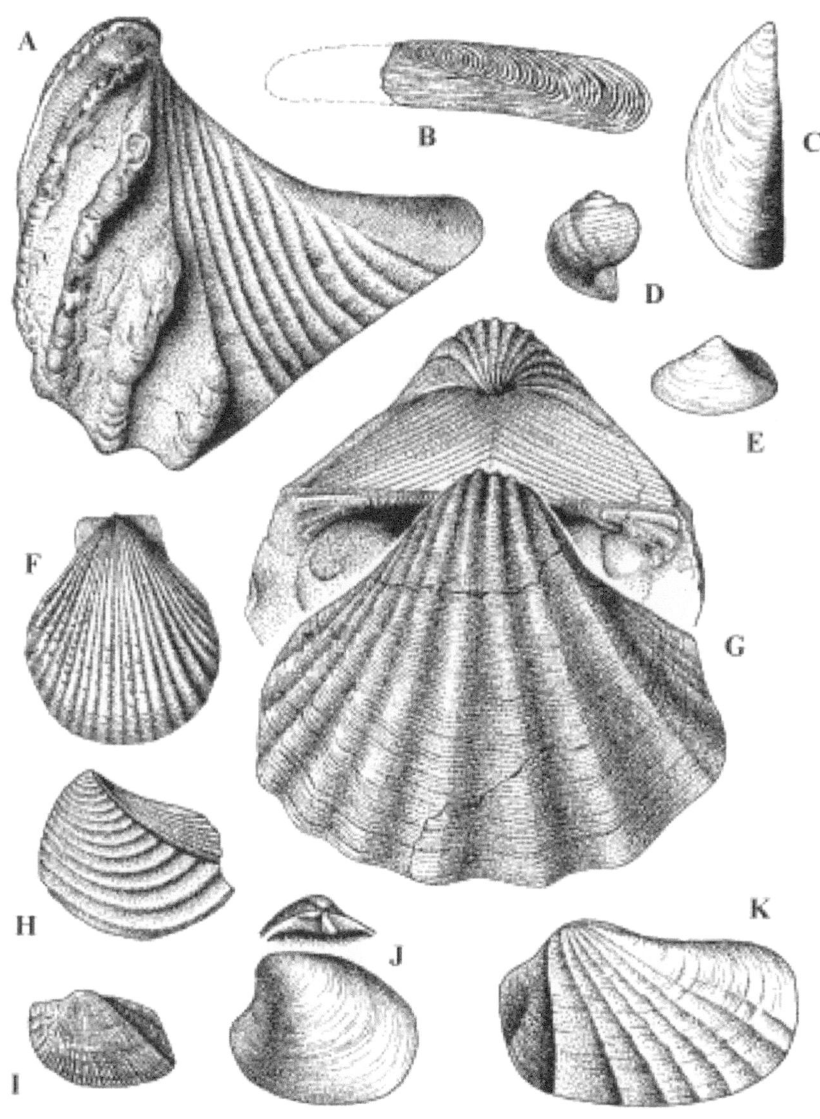

Plate 4

Plate 5

Late Valanginian

Asteroid, bivalve, plesiosaur and ammonite
from the Sundays River Formation

A. *Ophiolancea swartkopensis* Shone, x2.
B. *Arcomya baini* (Sharpe).
C. *Leptocleidus capensis* (Andrews), x0.3.
D-E. *Olcostephanus atherstonei* (Sharpe), x0.5.

All x1 unless stated otherwise.

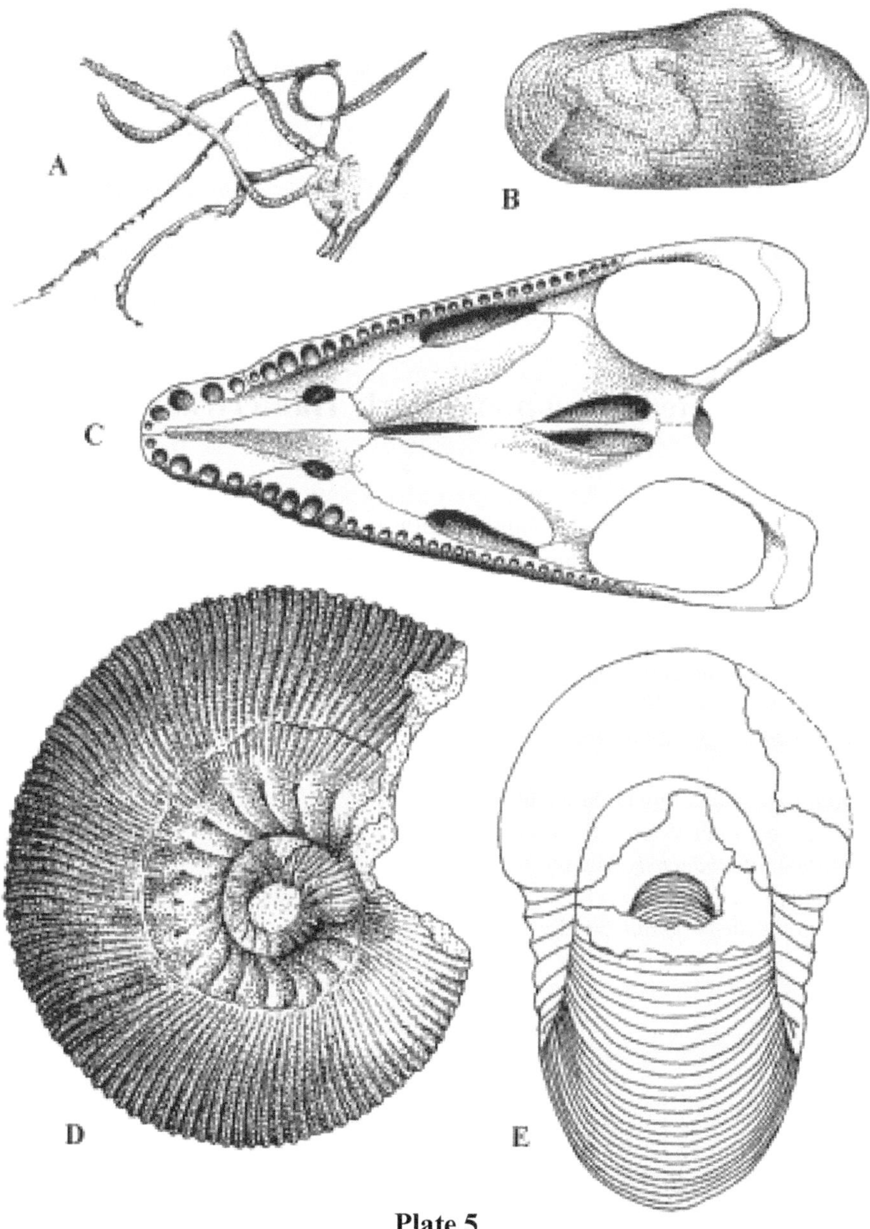

Plate 5

Plate 6

Late Valanginian - Early Hauterivian

Bivalves and ammonites from the
Sundays River Formation

A. *Mytiloperna atherstoni* (Sharpe).
B. *Seebachia bronni* (Krauss).
C. *Bochianites africanus* (Tate).
D. *Phyllopachyceras rogersi* (Kitchin), x2.
E. *Entolium orbiculare* (J. Sowerby).
F. *Rinetrigonia ventricosa* (Krauss).
G. *Natica uitenhagensis* Kitchin.
H. *Camptonectes cottaldinus* (d'Orbigny).
I. *Neohoploceras subanceps* (Tate), x2.

All x1 unless stated otherwise.

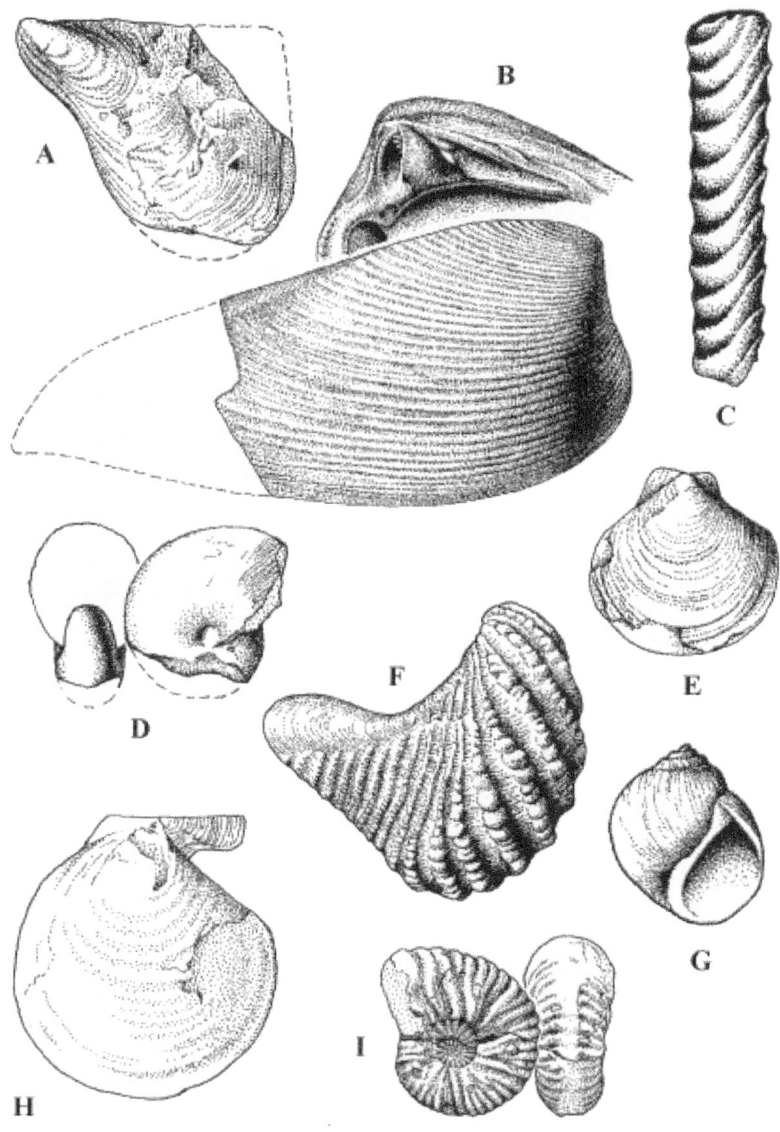

Plate 6

Plate 7

Late Valanginian

Gastropods, bivalves and ammonite
from the Sundays River Formation

A. *"Turbo" minulutus* (Kitchin), x4.
B. *Thetis papyracea* (Sharpe).
C. *Mayesella vau* (Kitchin).
D. *Actaeonina atherstoni* (Sharpe), x3.
E. *Acesta obliquissima* (Tate).
F. *Austromyophorella oosthuizeni* (Cooper).
G. *Mactra? dubia* Kitchin, x1.5.
H. *Pterotrigonioides rogersi* (Kitchin).
I. *Meretrix uitenhagensis* Kitchin, x1.5.
J. *Anthonya lineata* Kitchin, x2.
K. *Miodesmoceras haughtoni* (Spath), x2.

All x1 unless stated otherwise.

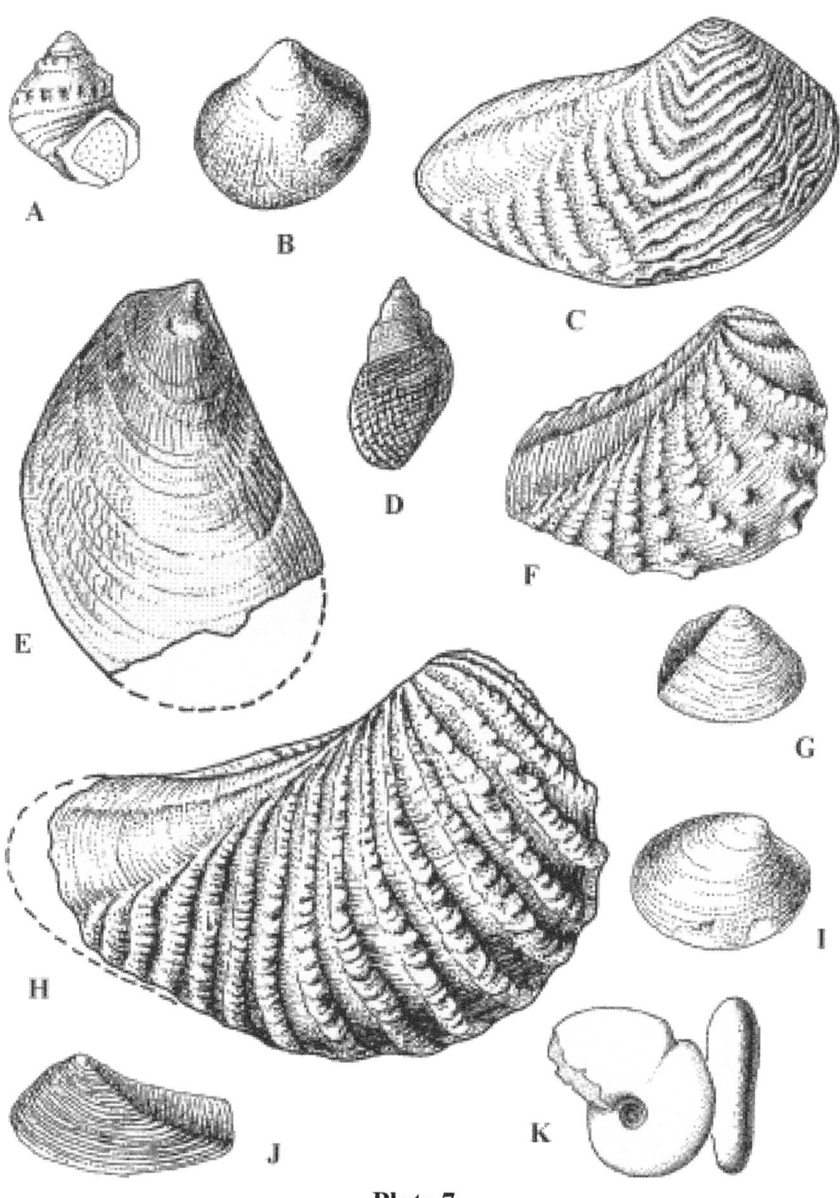

Plate 7

Plate 8

Late Valanginian

Bivalve from the Sundays River Formation

Transitrigonia herzogi (Goldfuss), x1.

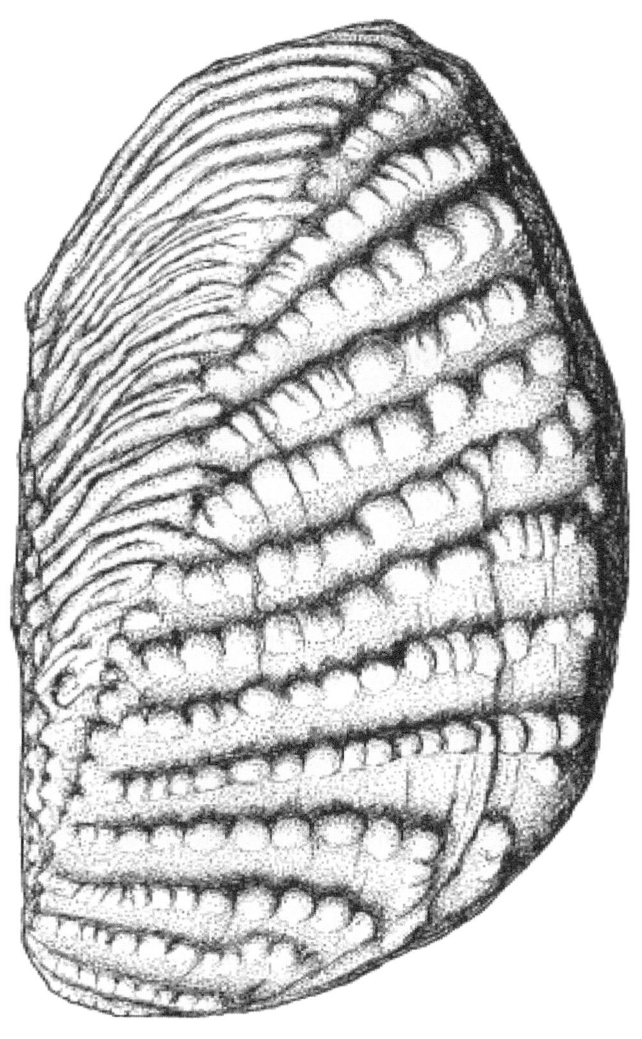

Plate 8

Plate 9

Late Valanginian - Early Hauterivian

Ammonites and gastropod from the
Sundays River and Kikundi Formations

A-B. *Criosarasinella spinosissimum* (Hausmann), x0.67.
C. *Olcostephanus baini* (Sharpe) (a microconch).
D. *Jeannoticeras jeannoti* (d'Orbigny), x0.8.
E. *Olcostephanus multistriatus* (Zwierzycki), x0.75.
F. *Bathrotomaria uitenhagensis* (Rennie), x0.75.

All x1 unless stated otherwise.

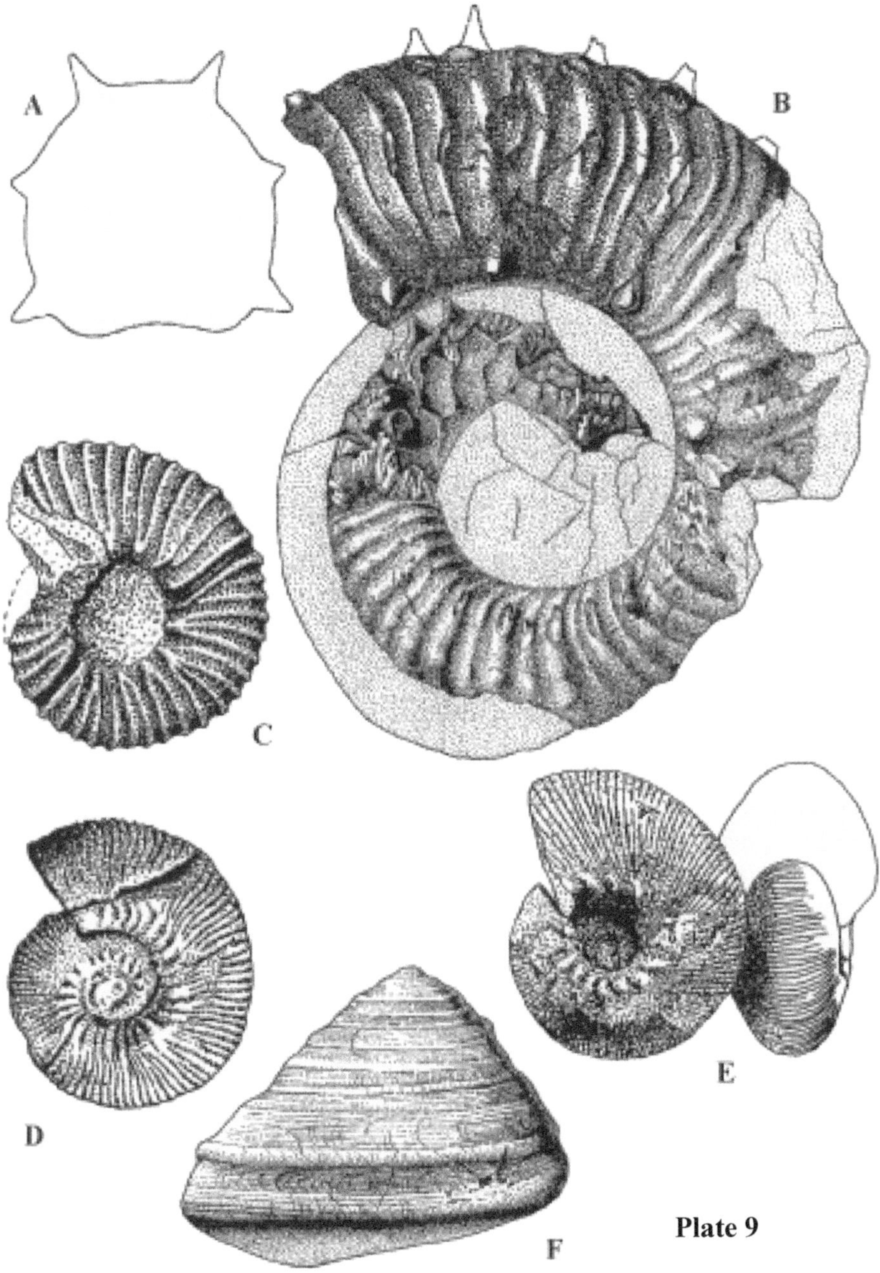

Plate 9

Plate 10

Late Valanginian

Bivalves, ammonite and belemnite from the
Sundays River and Brenton Formations

A. *Megatrigonia conocardiiformis* (Krauss).
B. *Hibolites* cf. *subfusiformis* (Raspail).
C. *Pterotrigonioides savagei* (Cooper), x2.
D. "*Hybonoticeras*" sp.
E. *Isognomon brentonensis* (Schwarz).
F. *Isognomon theseni* (Schwarz).
G. *Notoscabrotrigonia kitchini* (Schwarz).
H. *Neuquenella kensleyi* (Cooper), a juvenile..

All x1 unless stated otherwise.

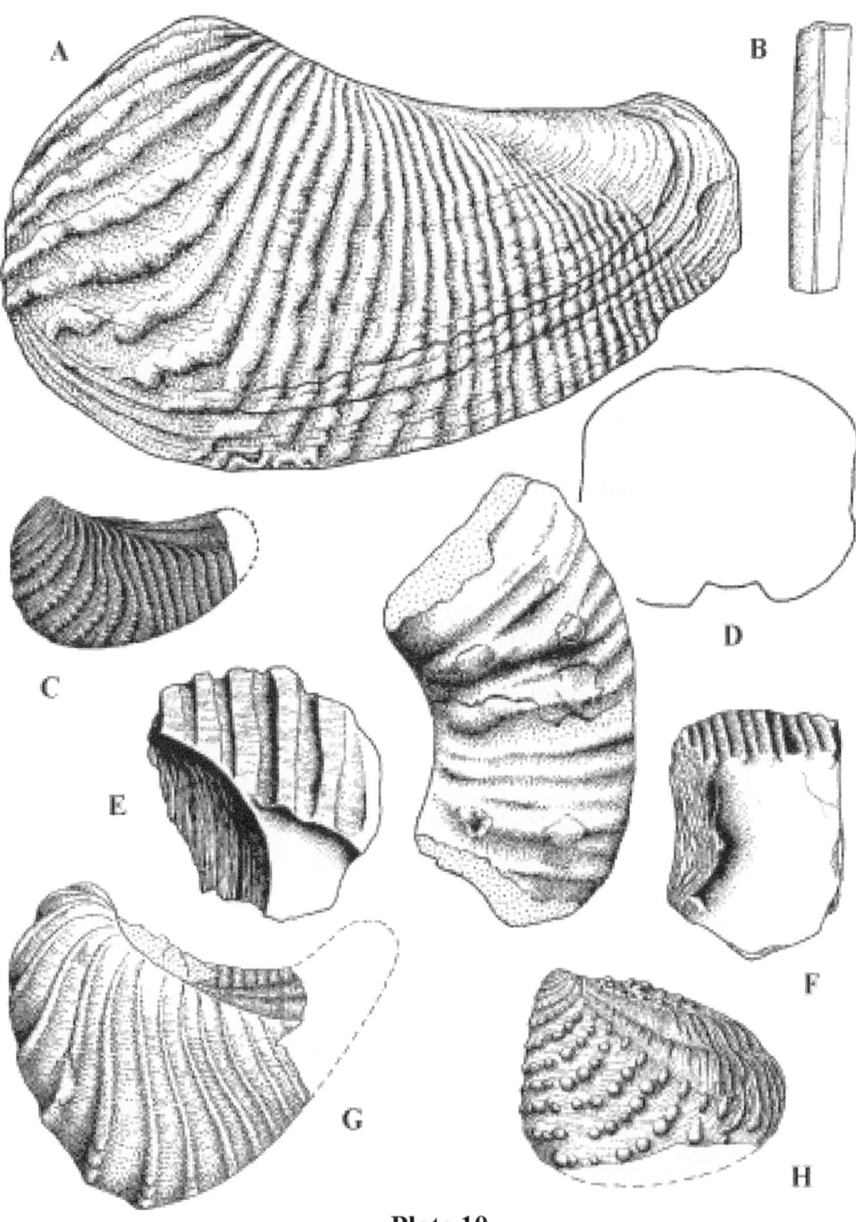

Plate 10

Plate 11

Barremian

Bivalves and gastropod from the Makhatini Formation

A. *Pseudoyaadia hennigi* (Lange).
B. *Austromyophorella dooleyi* Cooper, x1.5.
C. *Confusiscala borgesi* (Rennie), x3.
D. *Mayesella haughtoni* (Rennie).
E-F. *Utrobiqueostreon greylingae* Cooper, x2.

All fossils x1, unless stated otherwise.

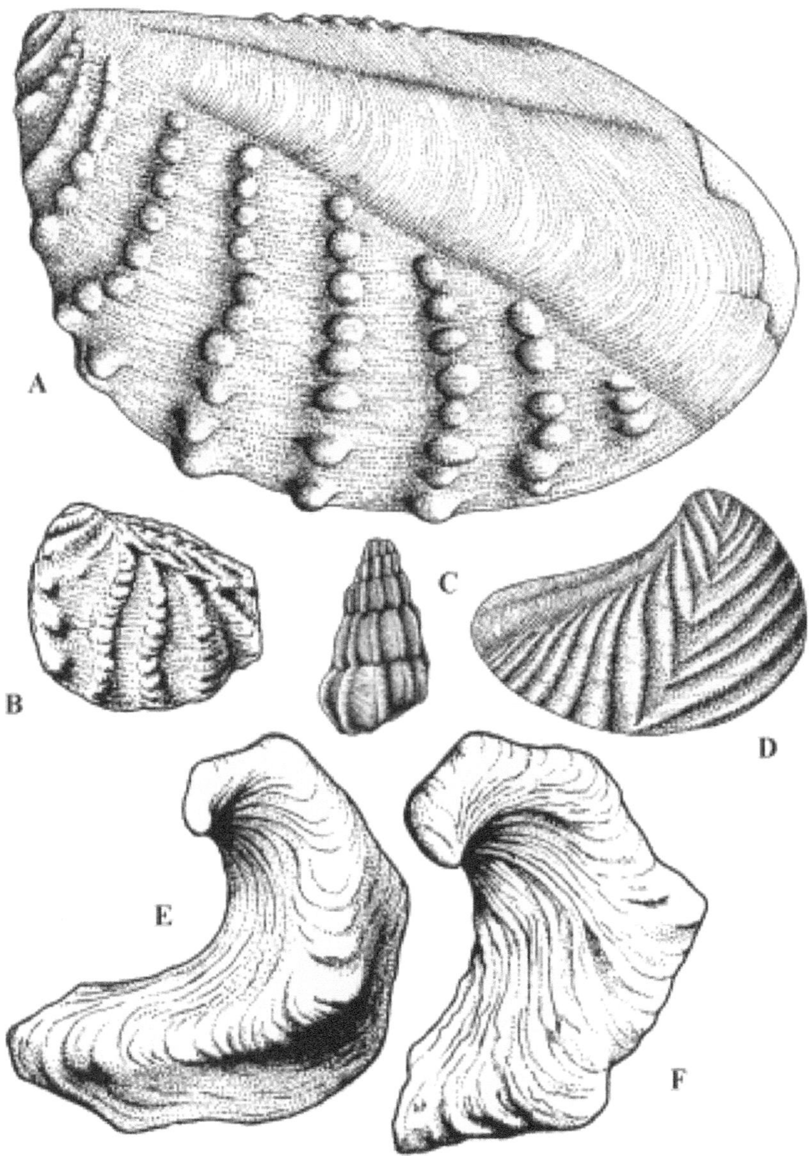

Plate 11

Plate 12

Barremian

Bivalves, ammonite and gastropods from the Makhatini Formation

A. *Coelastarte coxi* (Rennie), x1.5.
B. *Cryptocrioceras yrigoyeni* (Leanza), x0.5 (after Klinger & Kennedy).
C. *Gyrodes* cf. *genti* (Sowerby), x1.5.
D. *Confusiscala chalalensis* (Rennie), x1.5.
E. *Aetostreon latissimum* (Lamarck).

All x1 unless stated otherwise.

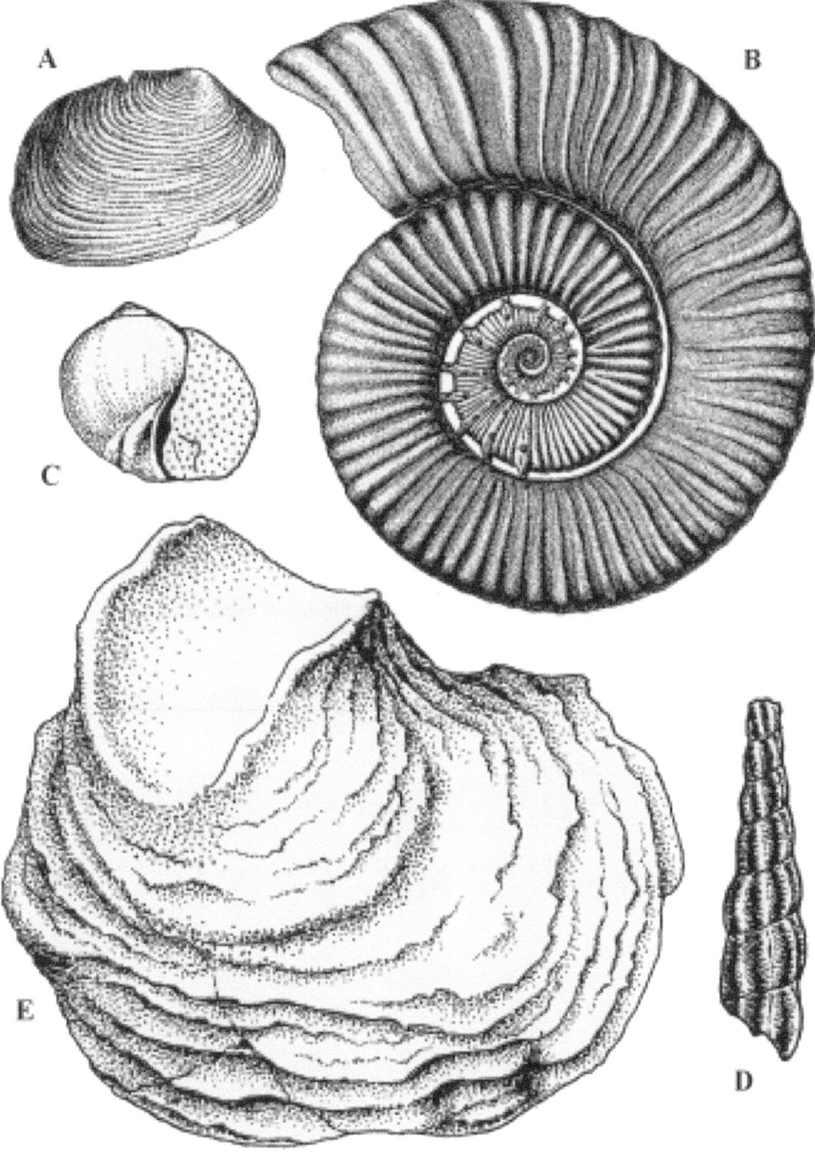

Plate 12

Plate 13

Barremian

Ammonites, bivalve and belemnite from the Makhatini Formation

A. *Colchidites vulanensis australis* Klinger, Kakabadze & Kennedy.
B. *Trigonia* cf. *tatei* Neumayr.
C. *Sanmartinoceras africanum* Kennedy & Klinger.
D. *Acrioceras zulu* Klinger & Kennedy, x1.5 (after Klinger & Kennedy).
E. *Heteroceras elegans* Rouchadzé.
F. *Chalalabelus renniei* (Spath).

All x1 unless stated otherwise.

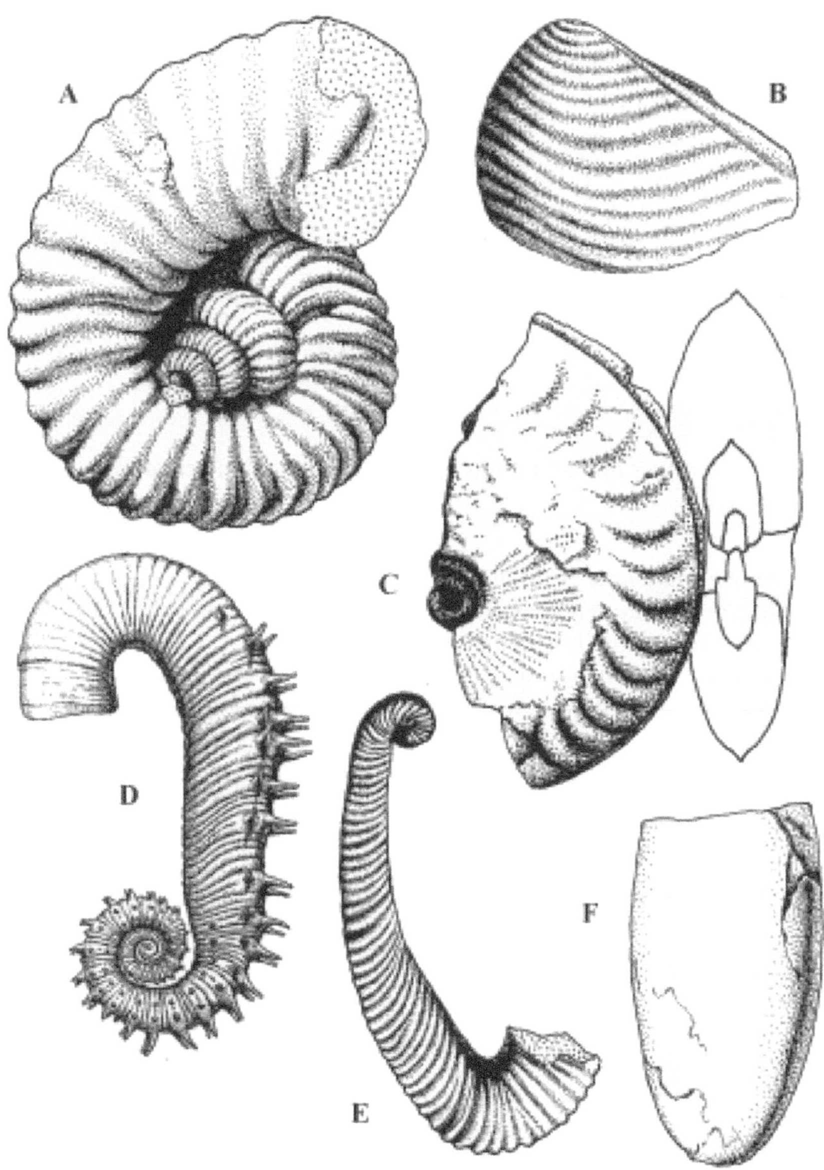

Plate 13

Plate 14

Early Aptian

Bivalves and ammonites from the Niongala Formation

A. *Tanzanitrigonia schwarzi* (Müller), x0.75.
B. *Proveniella? rothpletzi* (Krenkel).
C. *Noramya gabrielis* (d'Orbigny), x0.75.
D. *Neithea atava* (d'Orbigny), x0.75 (syn. *Vola lindiensis* Lange).
E. *Sphaera corrugata* J. Sowerby, x0.75 (syn. *Venus cordiformis* Deshayes).
F. *Vectianella? tellina* (Lange).
G. *Ptychomya plana* Agassiz (syn. *P. kitchini* Lange).

All x1 unless stated otherwise.

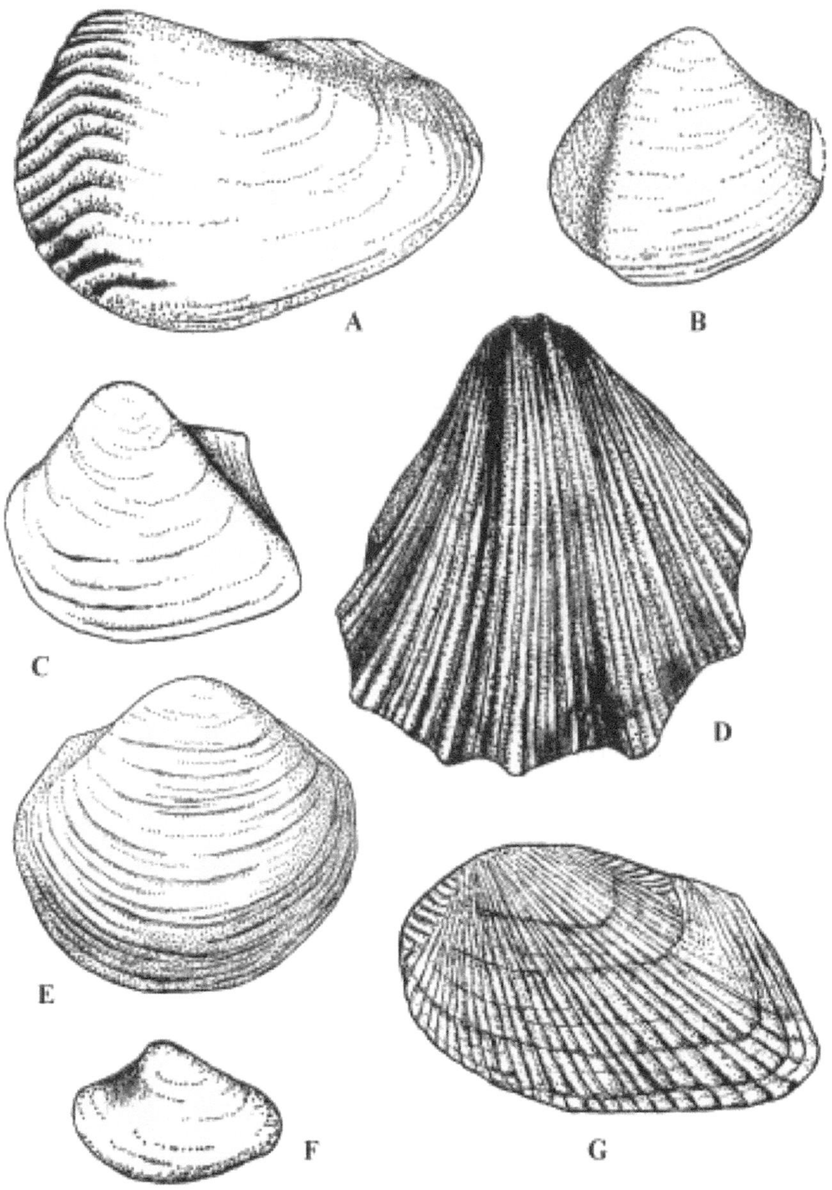

Plate 14

Plate 15

Barremian - Early Aptian

Bivalves , ammonites and belemnite from the Makhatini Formation

A. *Megatrigonia saggersoni* (Cooper).
B. *Zulutrigonia pongolensis* (Rennie).
C. *Audouliceras* cf. *ajax* (Anderson), x0.5.
D. *Cheloniceras gottschei* (Kilian).
E. *Peratobelus foersteri* Doyle.
F. *Pratulum rogersi* (Rennie).

All x1 unless stated otherwise.

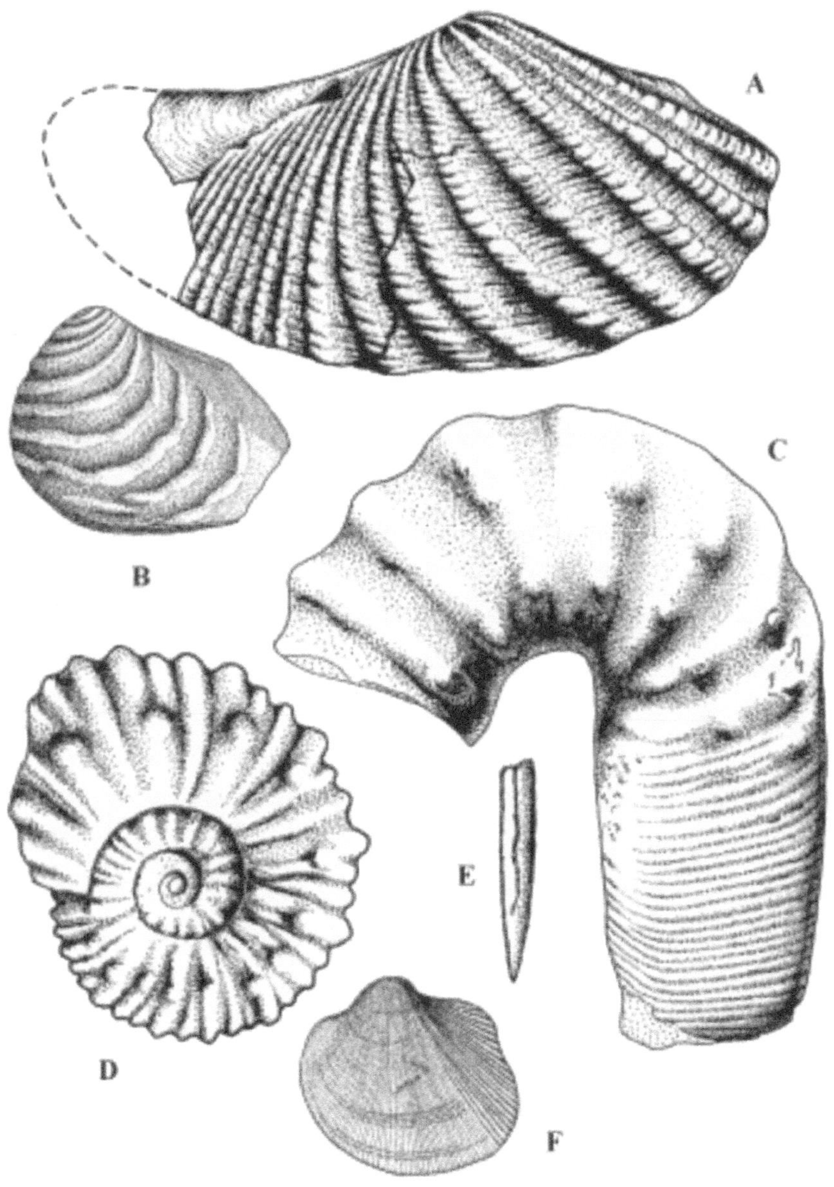

Plate 15

Plate 16

Late Aptian

Bivalves from the Ndabana Formation

A. *Zaletrigonia inconstans* (van Hoepen), x0.67.
B. *Pisotrigonia salebrosa* van Hoepen.
C. *Gervillella dentata* (Krauss).
D. *Isognomon ricordeanum* (d'Orbigny).

All x1 unless stated otherwise.

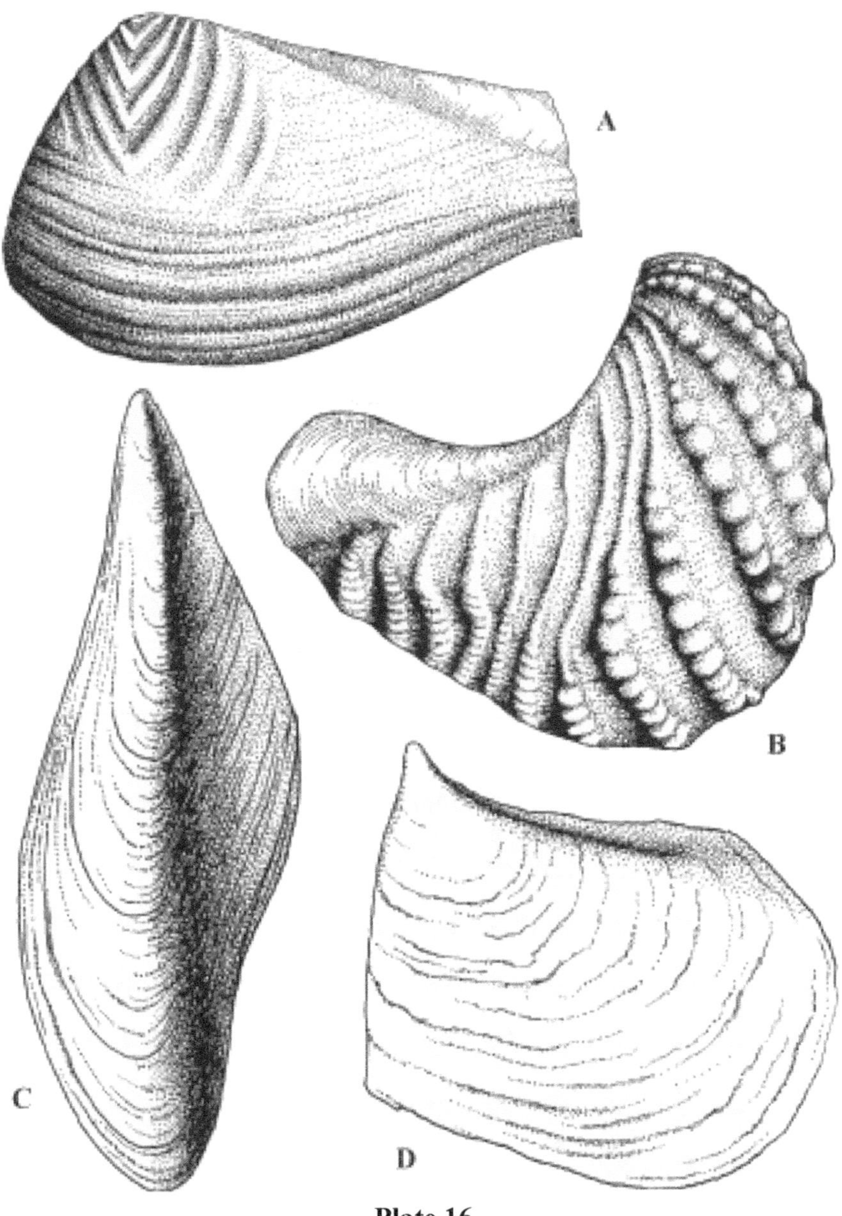

Plate 16

Plate 17

Late Aptian

Ammonites and bivalves from the
Ndabana and Maputo Formations

A. *Nolaniceras uhligi* (Anthula), x0.5.
B. *Nototrigonia shonei* Cooper, x1.5.
C. *Aconeceras nisus (*d'Orbigny).
D. *Acanthohoplites bigoureti* (Seunes).
E. *Panopea gurgitis* (Goldfuss).
F. *Calva* cf. *subrotunda* (J. de C. Sowerby).
G. *Megatrigonia obesa* van Hoepen.

All x1 unless stated otherwise.

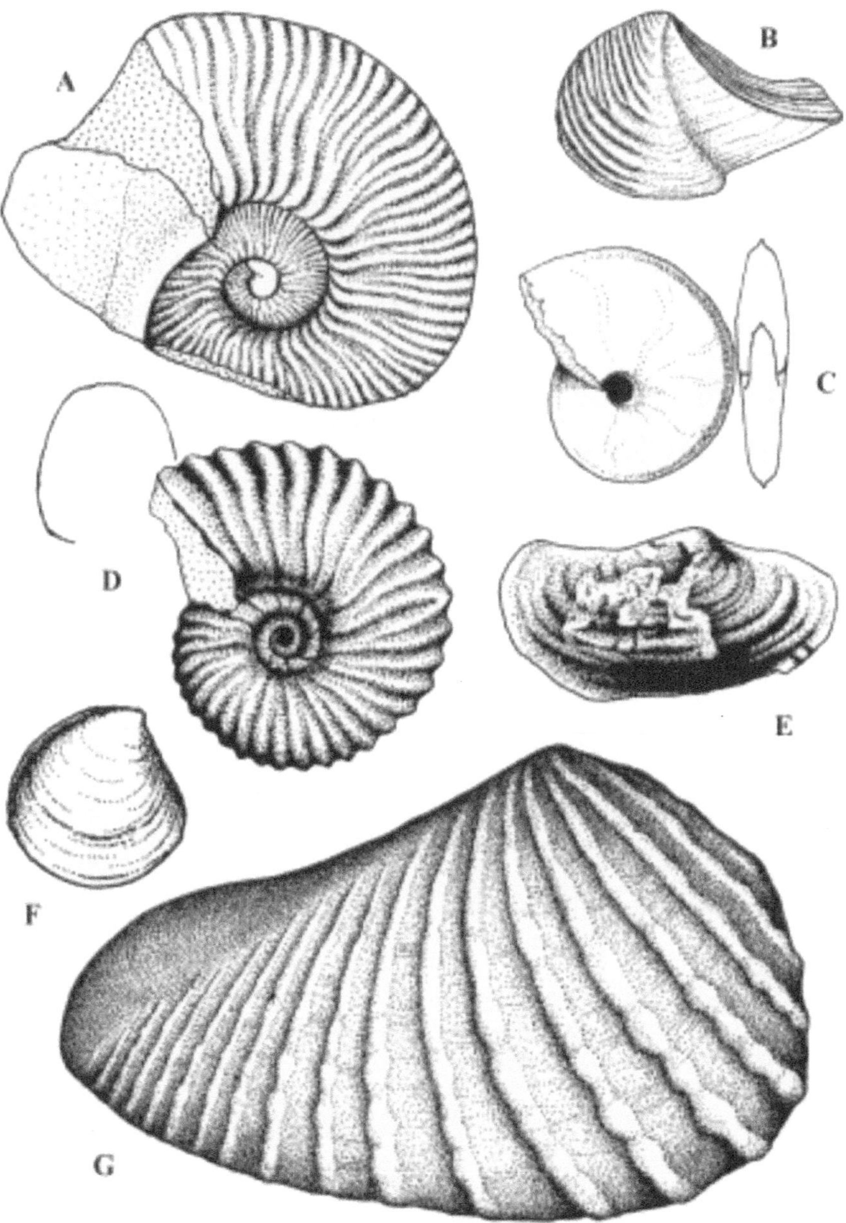

Plate 17

Plate 18

Late Aptian

Ammonites, bivalves and brachiopods from the Ndabana Formation and the Formação com *Pholadomya*

A. *Sinzovia trautscholdi* (Sinzow).
B. *Pholadomya pleuromyaeformis* Choffat, x1.5.
C. *Proveniella* cf. *meyeri* (Woods).
D. *Diadochoceras nodosocostatum* (d'Orbigny), x1.3.
E. *Acanthohoplites aschiltaensis* (Anthula).
F. *Helicancyloceras crassetuberculatum* Klinger & Kennedy.
G. *Cyclothyris* sp., x2
H. Undescribed spiriferoid brachiopod.

All x1unless stated otherwise.

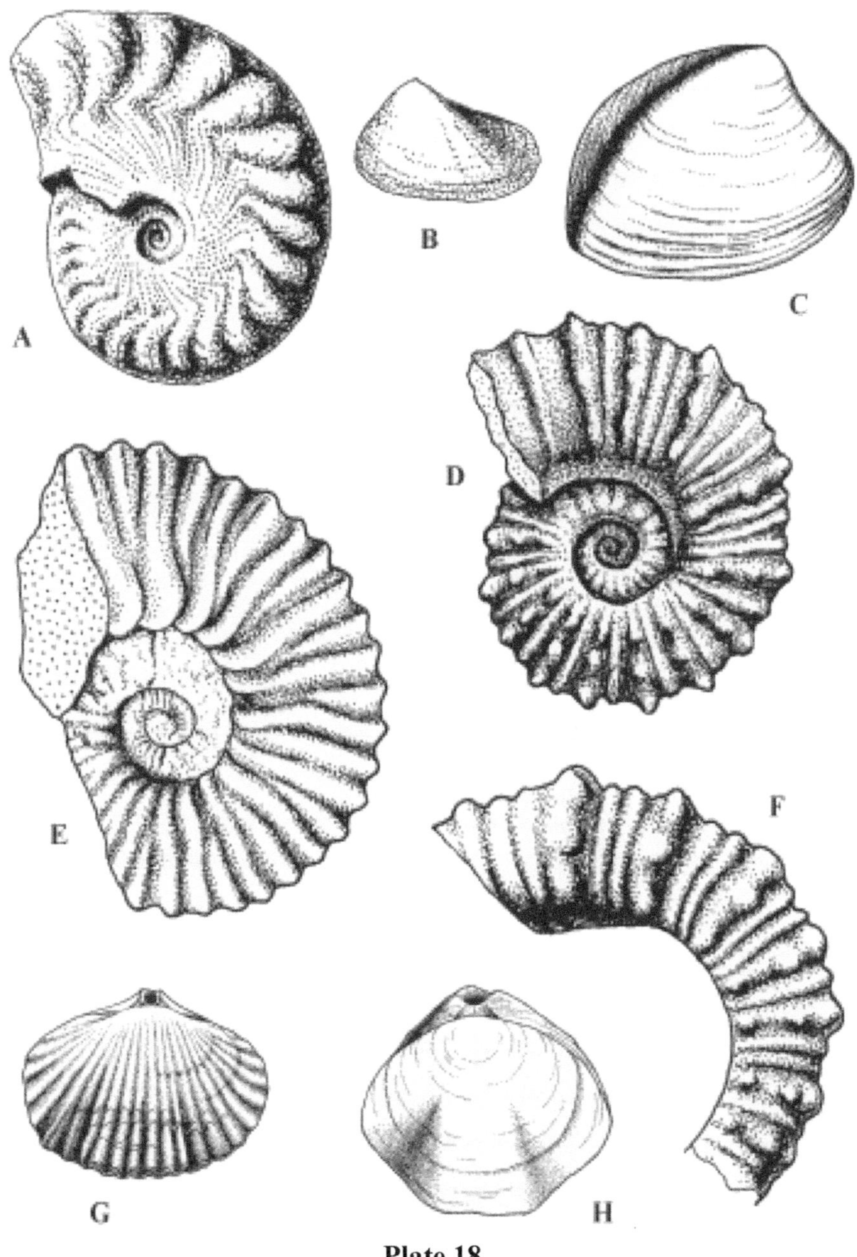

Plate 18

Plate 19

Early - Middle Albian

Ammonites and bivalves from the Mzinene Formation

A. *Tegoceras camatteanum* (d'Orbigny).
B. *Neithea syriaca* (Conrad), x2.
C. *Ptilotrigonia lauta* van Hoepen.
D. *Actinoceramus salomoni* (d'Orbigny).
E. *Pseudavicula? africana* Etheridge, x2.
F. *Pseudhelicoceras catenatum* (d'Orbigny).
G. *Sphenotrigonia frommurzei* (Rennie). This species ranges down into the Late Aptian.

All x1 unless stated otherwise.

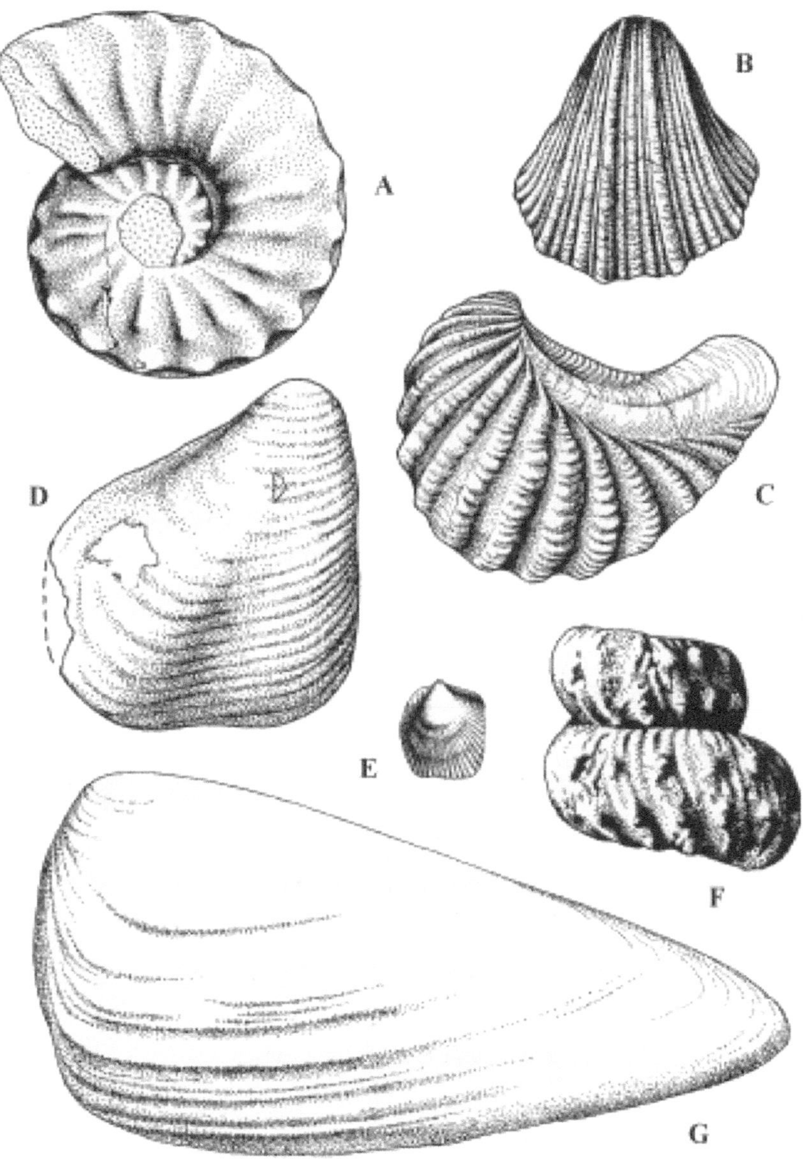

Plate 19

Plate 20

Middle - Late Albian

Brachiopod, bivalves, ammonites and gastropod
from the Mzinene Formation

A. *Dzirulina haughtoni* Owen, x2.
B. *Pterotrigonia cristata* van Hoepen.
C. *Idonearca? umsinenensis* (Etheridge).
D. *Douvilleiceras mammillatum aequinodum* (Quenstedt), x0.5.
E. *Alopecoceras ankeritterae* Kennedy & Klinger, a juvenile.
F. *Latiala bailyi* (Etheridge).
G. *Mesocallista andersoni* (Newton), x1.25.
H. *Epicyprina sanctaeluciensis* (Etheridge), x0.67.
I. *Globocardium sphaeroideum* (Forbes), x0.75.

All x1 unless stated otherwise.

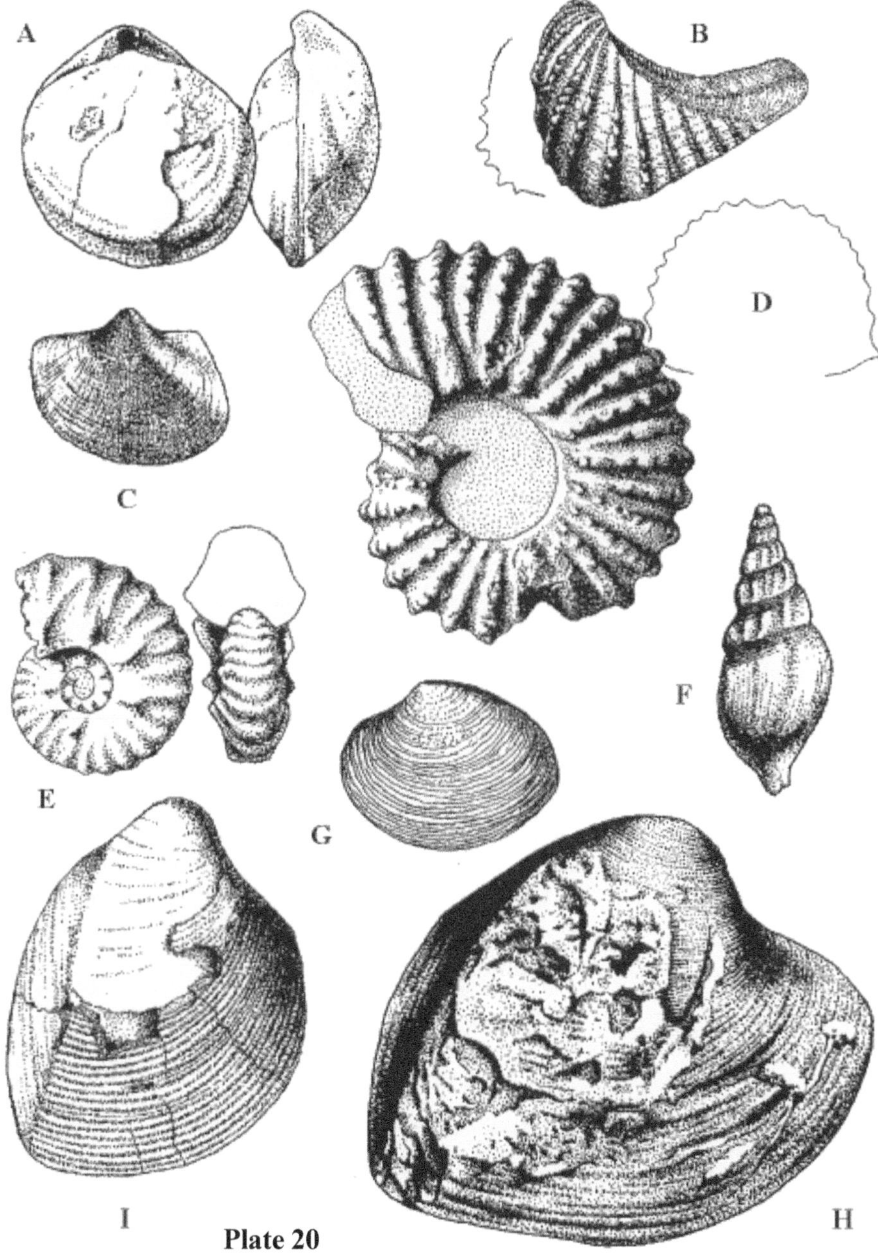

Plate 20

Plate 21

Early Late Albian

Ammonites, bivalves and gastropod
from the Mzinene Formation

A. *Rinetrigonia laevicosta* (Cooper).
B. *Manuaniceras manuanense* (Spath).
C. *Solariella africana* (Newton), x2.
D. *Rinetrigonia maccarthyi* (Cooper).
E. *Amphidonte arduennense austroafricanum* Cooper.
F. *Dipoloceras cristatum* (Deluc).
G. *Ptilotrigonia cricki* (Newton).

All x1 unless stated otherwise.

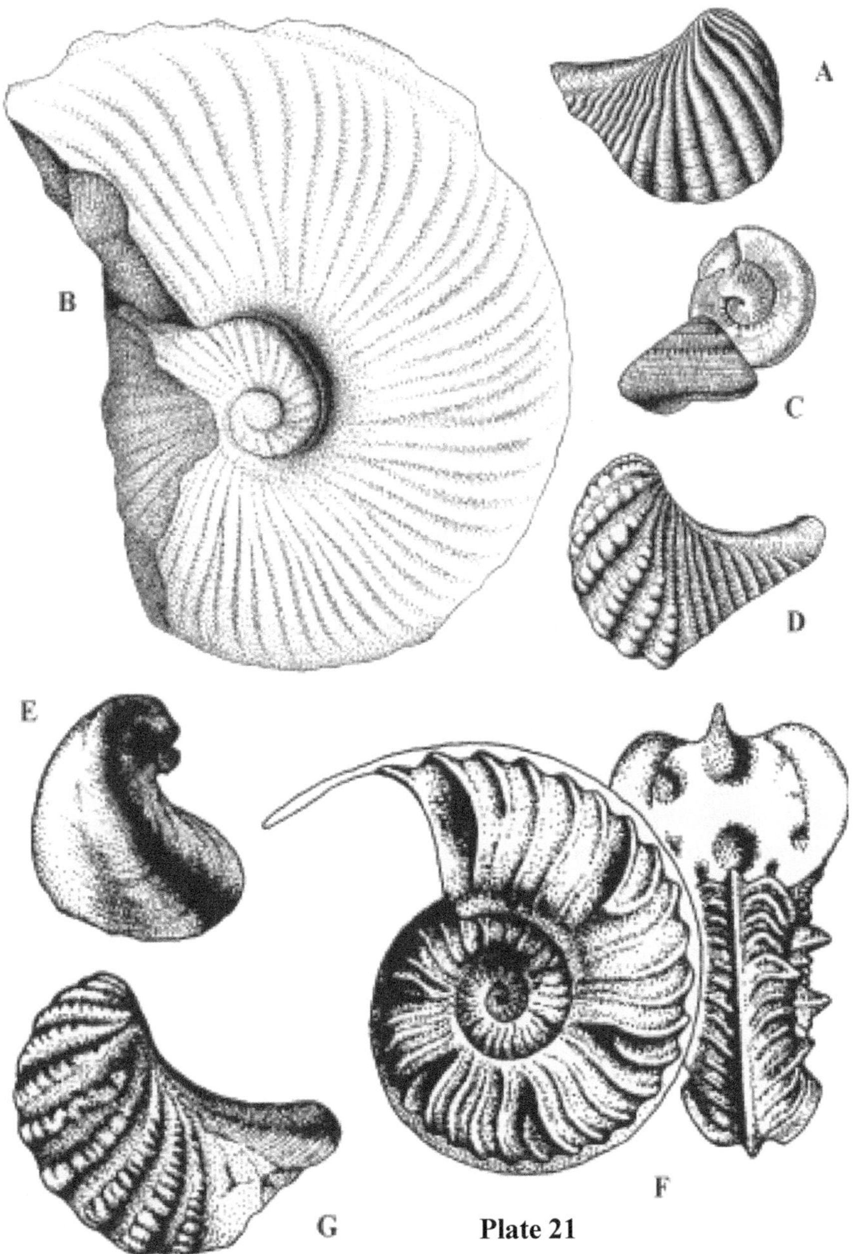

Plate 21

Plate 22

Late Albian

Ammonites, bivalves and gastropods from the Mzinene Formation

A. *Poikiloceras firmum* (van Hoepen).
B. *Lethargoceras incommodum* van Hoepen.
C. *Rutitrigonia peregrina* van Hoepen.
D. *Ptilotrigonia lautissima* van Hoepen.
E. *Jauberticeras jauberti* (d'Orbigny).
F. *Actinoceramus sulcatus* (Parkinson).
G. *Puzosia provincialis* (Parona & Bonarelli) - this genus is common, although this species has not yet been recorded.
H. *Glycymeris griesbachi* (Newton).
I. *Mesoglauconia madagascariensis* Mennessier.
J. *Fossarus* aff. *besairiei* Collignon.
K. *Plicatula andersoni* Newton (this species may be Coniacian).
L. *Eunatica* sp.
M. *Allomactra* sp. nov., x2.

All x1 unless stated otherwise.

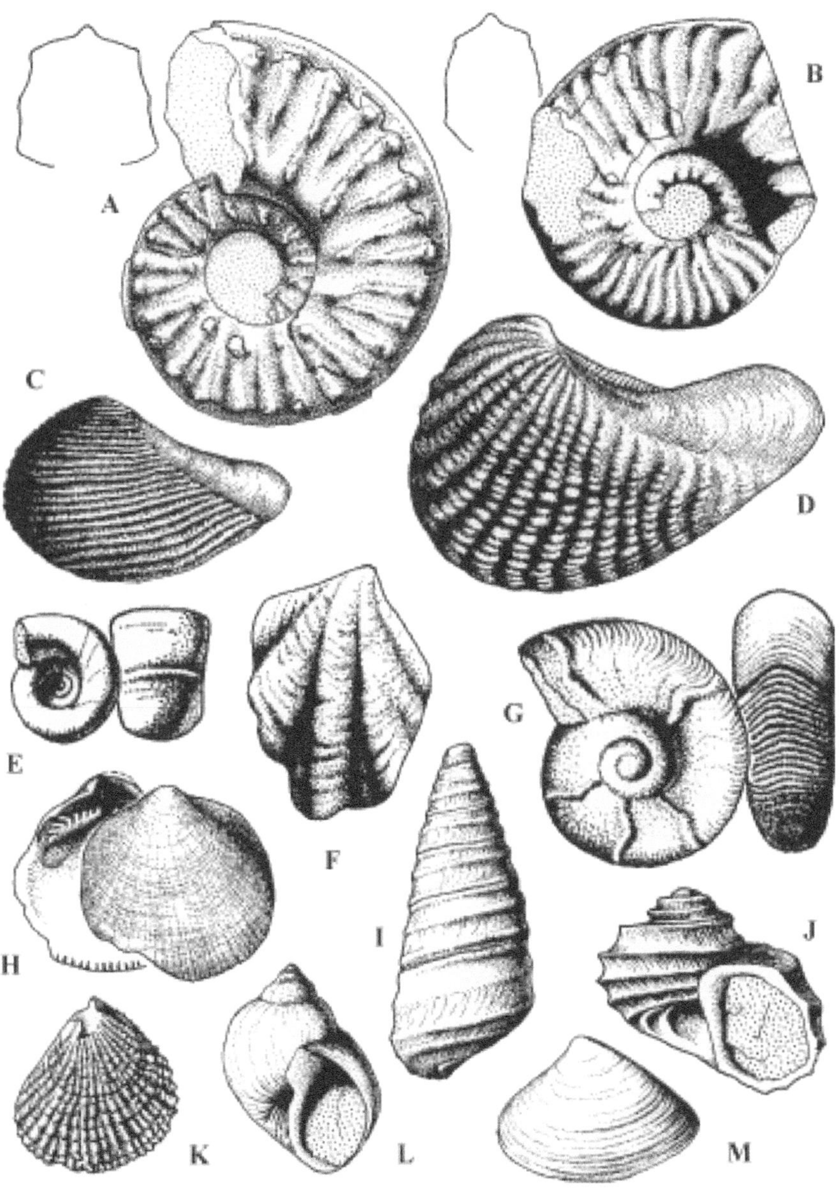

Plate 22

Plate 23

Late Albian

Ammonites, bivalves, echinoid and belemnite
from the Mzinene and Maputo Formations

A. *Arestoceras splendidum* van Hoepen.
B. *Hysteroceras adelei* van Hoepen.
C. *Plicatula rogersi* Newton.
D. *Hysteroceras carinatum* Spath.
E. *Styphloceras latidorsatum* van Hoepen.
F. *Cainoceras strigosum* van Hoepen.
G. *Neohibolites ewaldi* (von Strombeck).
H. *Pseudholaster vanhoepeni* (Besairie & Lambert).
I. *Proveniella* sp.
J. *Procardia vignesi* (Lartet).

All x1.

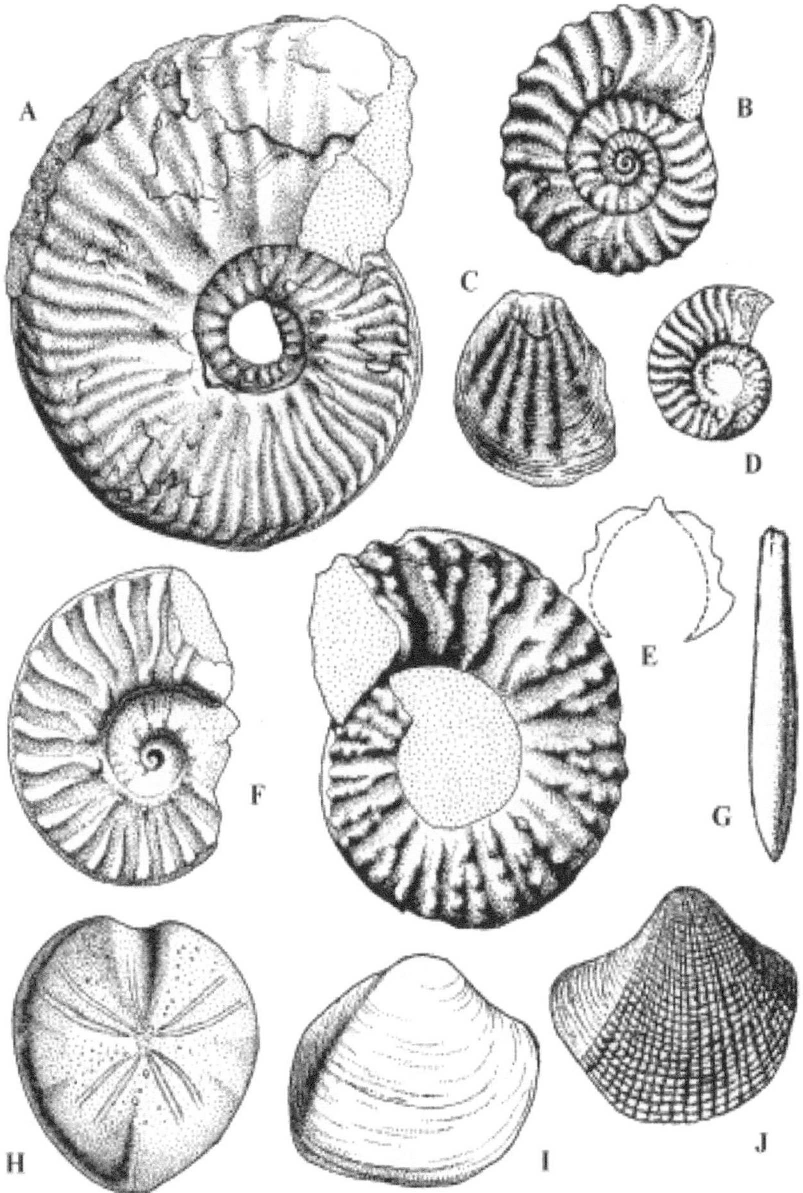

Plate 23

Plate 24

Late Albian

Ammonites and bivalves from the Mzinene Formation

A. *Pervinquieria vallifera* (van Hoepen).
B. *Actinoceramus concentricus* (Parkinson).
C. *Rhynchostreon matthewsi* Cooper, x2.
D. *Tetragonites subtimotheanus* Wiedmann.
E. *Labeceras plasticum crassum* Spath.
F. *Eupsectroceras strigosum* van Hoepen, x0.75.
G. *Zuluscaphites orycteropus* van Hoepen.
H. *Anagaudryceras pulchrum* (Crick).

All x1 unless stated otherwise.

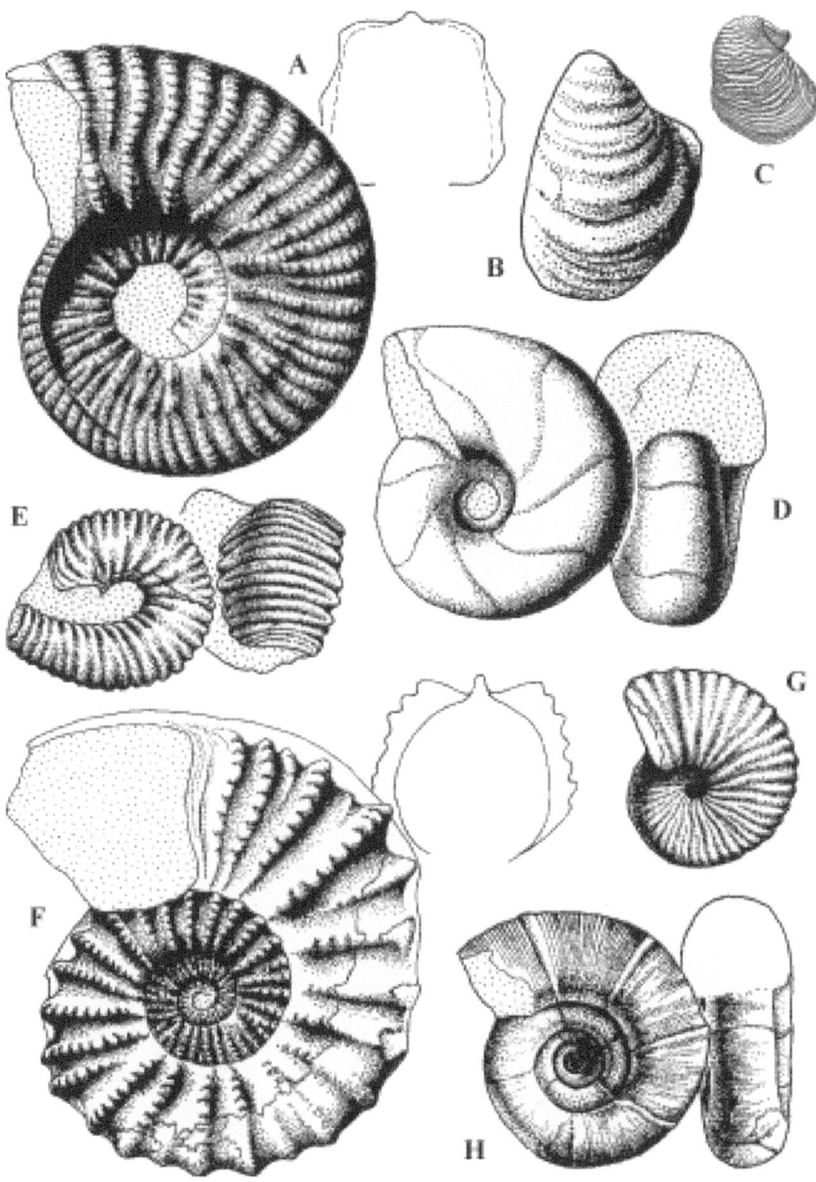

Plate 24

Plate 25

Late Albian

Nautiloids, ammonites, brachiopod and echinoid
from the Mzinene Formation

A. *Cymatoceras imbricatum* (Crick), x0.5.
B. *Paraturrilites circumtaeniatus* (Kossmat), x0.75.
C. *Cymatoceras manuanense* (Crick), x0.5.
D. *Praelongithyris vanhoepeni* (Lange), x1.3.
E. *Salaziceras salazense* (Breistroffer).
F. *Anisoceras perarmatum* Pictet & Campiche.
G. *Pygurus mendelssohni* Greyling & Cooper.

All x1 unless stated otherwise.

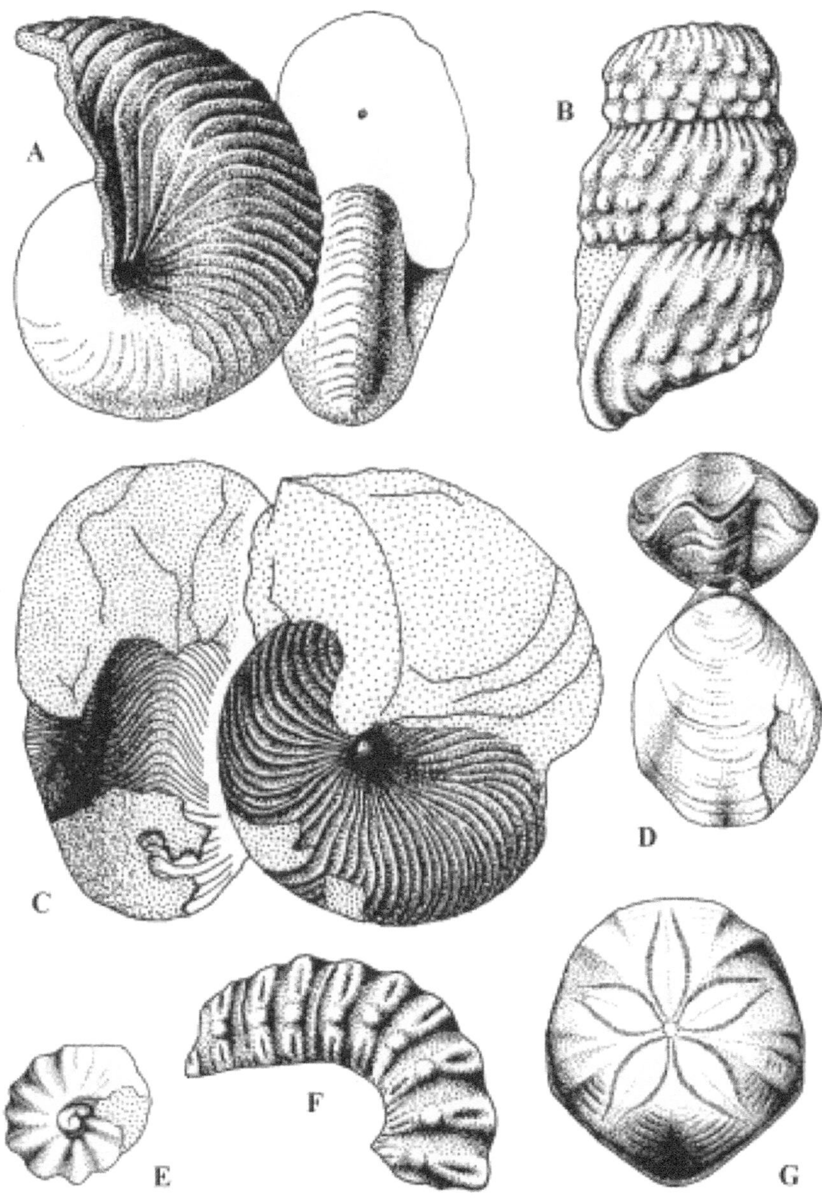

Plate 25

Plate 26

Late Albian

Ammonites and bivalves from the Mzinene Formation

A. *Ricnoceras pandai* van Hoepen.
B. *Actinoceramus volviumbonatus* (Etheridge).
C. *Amphidonte malchusi* Cooper, x0.8.
D. *Idonearca woodsi* (Crick).
E. *Deiradoceras prerostratum* (Spath), x0.4.
F. *Puzosia australis* Venzo, x0.4.
G. *Bhimaites spathi* (Venzo).

All x1 unless stated otherwise.

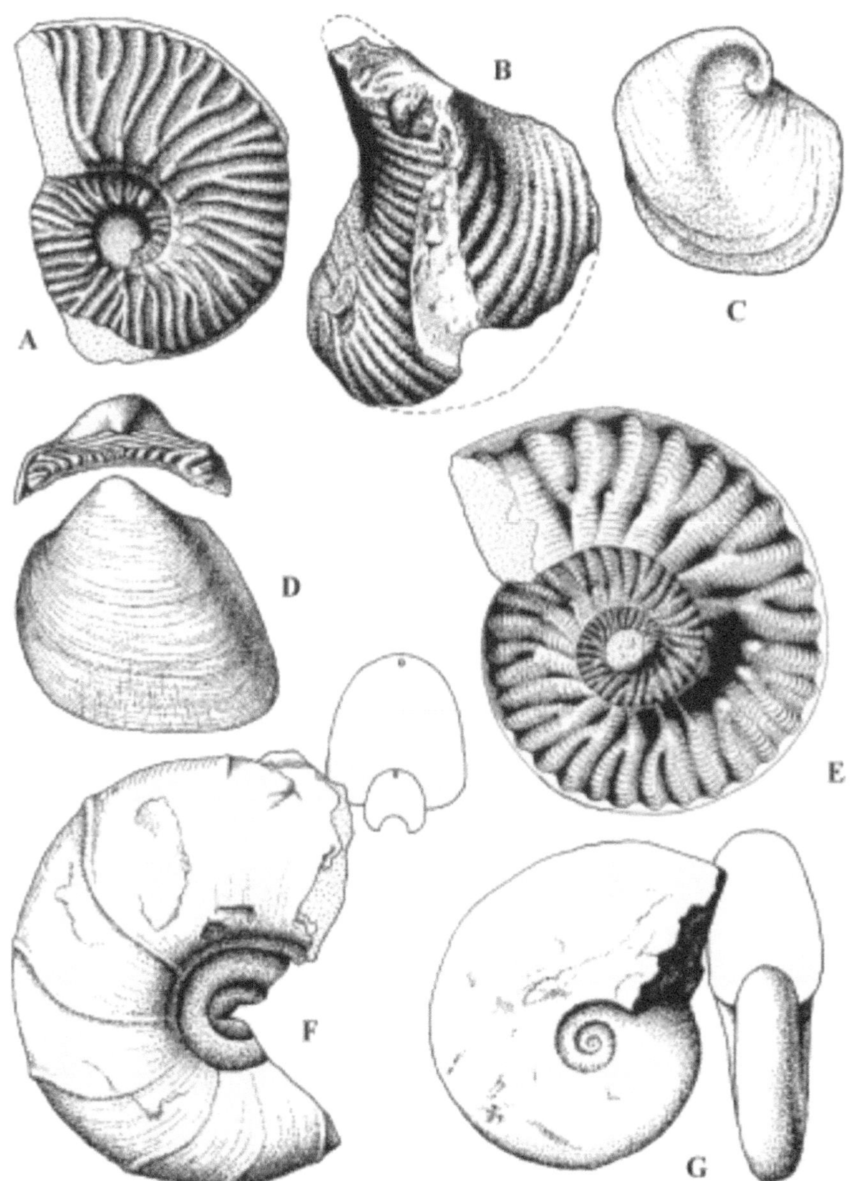

Plate 26

Plate 27

Late Albian

Ammonite from the Mzinene Formation

Achilleoceras erasmusi van Hoepen, x0.25.

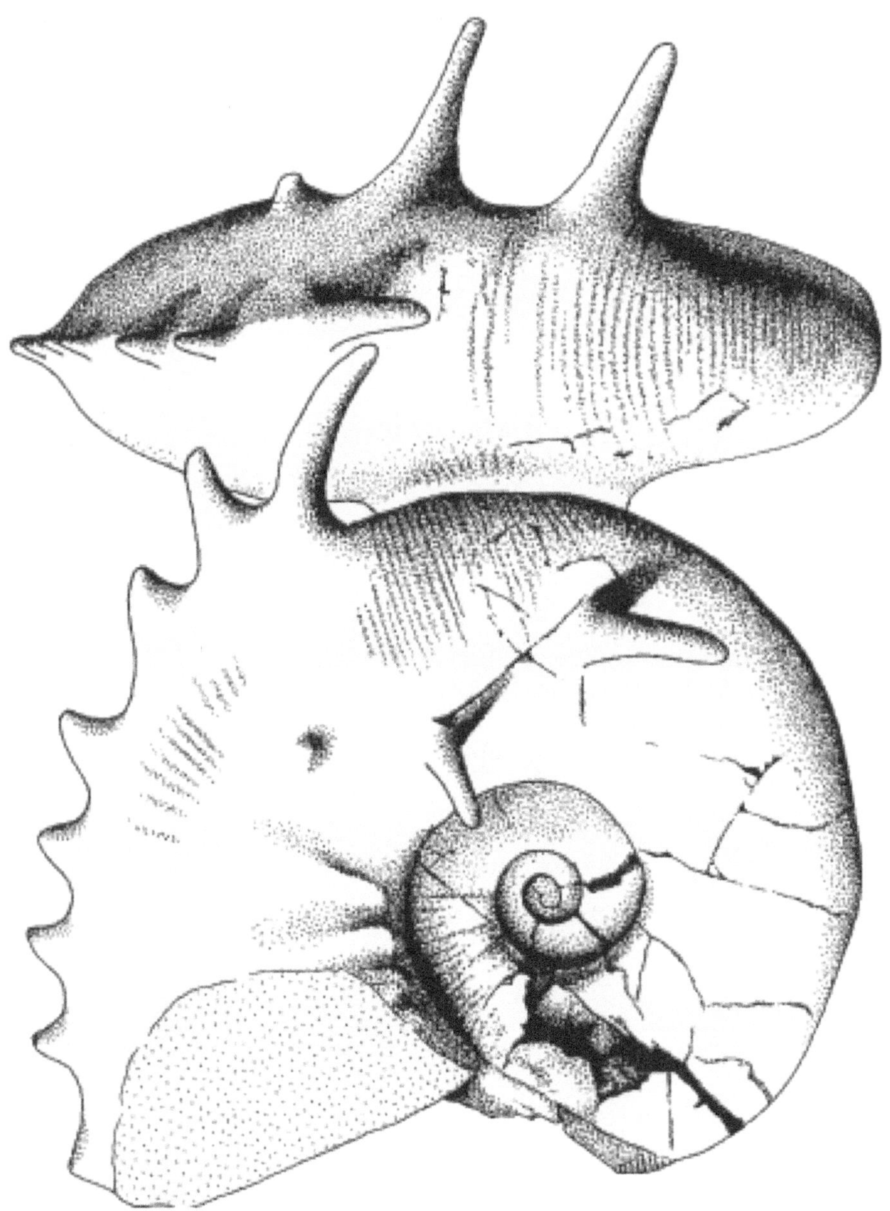

Plate 27

Plate 28

Late Albian

Ammonites, bivalves and gastropod from the Catumbella Formation

A-B. *Elobiceras elobiense* (Saznocha), x0.67.
C. *Drepanochilus reineckei* (Rennie), x1.5.
D. *Protocardia moutai* Rennie.
E-F. *Neokentroceras curvicornu* (Spath), x1.25.
G. *Lucina reineckei* Rennie, x0.8.
H. *Angolaites simplex* (Spath), x0.75.

All x1 unless stated otherwise.

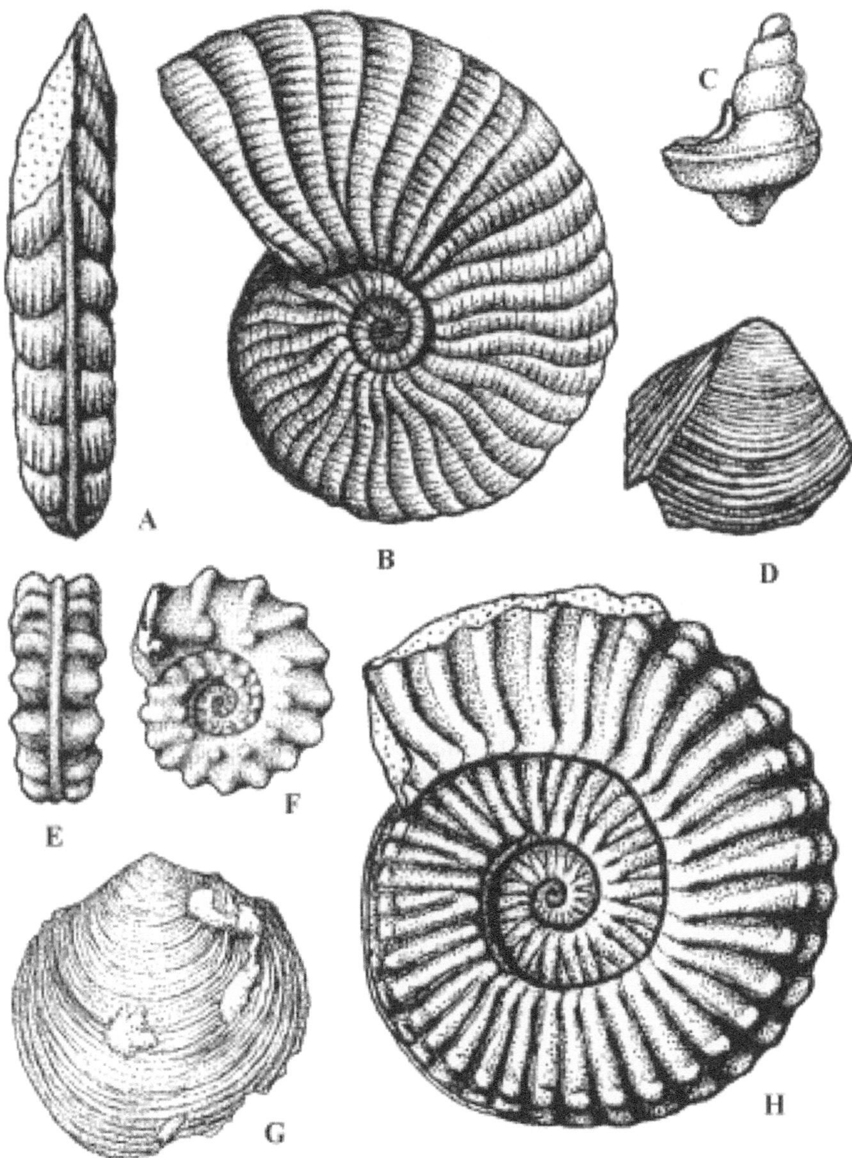

Plate 28

Plate 29

Late Albian

Gastropods, ammonites and echinoid
from the Cuio and Cabo Ledo Formations

A. *Actaeonella anchietai* Choffat, x0.8
B. *Prohysteroceras wordiei* Spath, x0.67.
C. *Eunerinea capelloi* (Choffat).
D. *Temnocidaris malheiroi* (de Loriol).
E. *Reyreiceras reyrei* Collignon, x0.7 (after Cooper & Kennedy 1979).
F-G. *Stoliczkaia tenuis* (Renz).

All x1 unless stated otherwise.

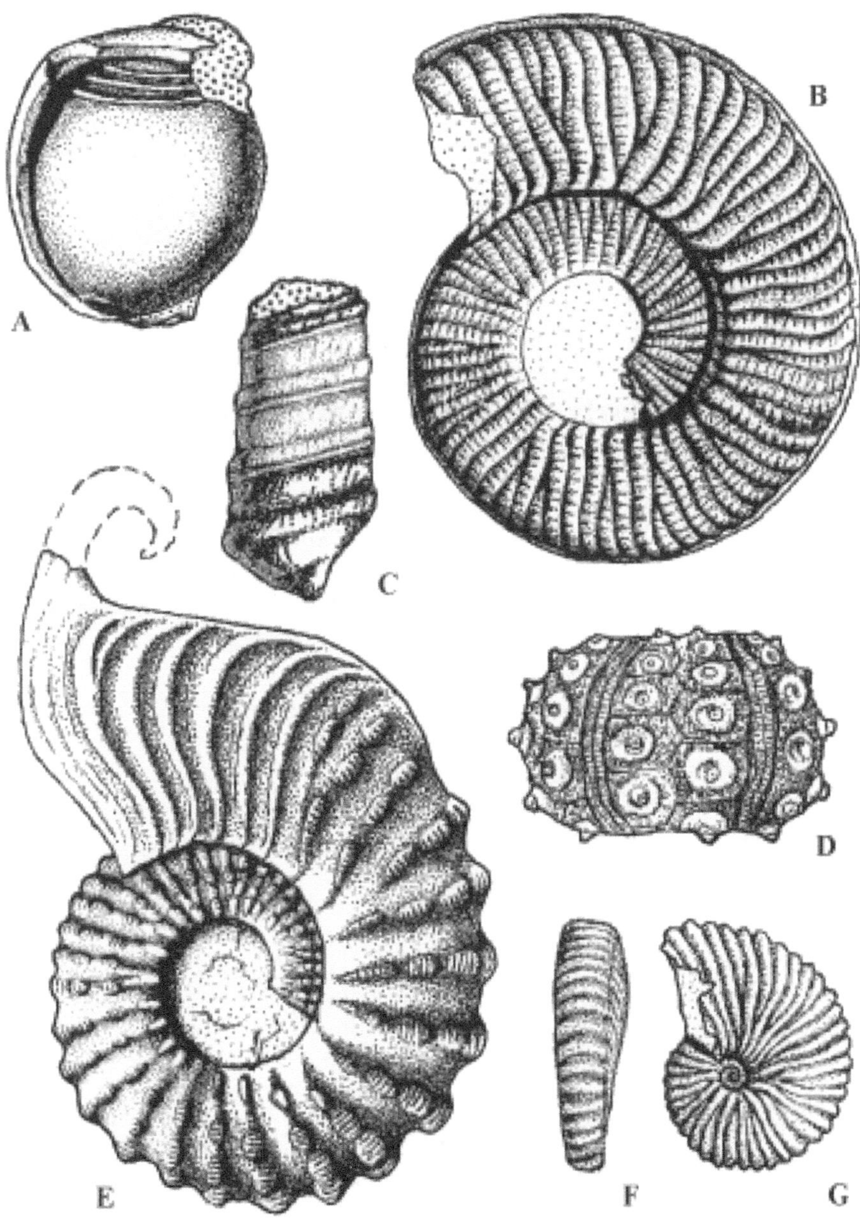

Plate 29

Plate 30

Early - Middle Cenomanian

Ammonites from the Mzinene and Maputo Formations

A. *Sharpeiceras florancae* Spath, x0.67.
B. *Turrilites costatus* Lamarck.
C. *Flickia quadrata* Collignon, x2.
D. *Neostlingoceras carcitanense* (Matheron).
E. *Mantelliceras saxbii* (Sharpe).
F. *Ostlingoceras rorayense* Collignon, x3.
G. *Mariella gallieni evoluta* Klinger & Kennedy, x2.
H. *Hypoturrilites cricki* Klinger & Kennedy.

All x1 unless indicated otherwise.

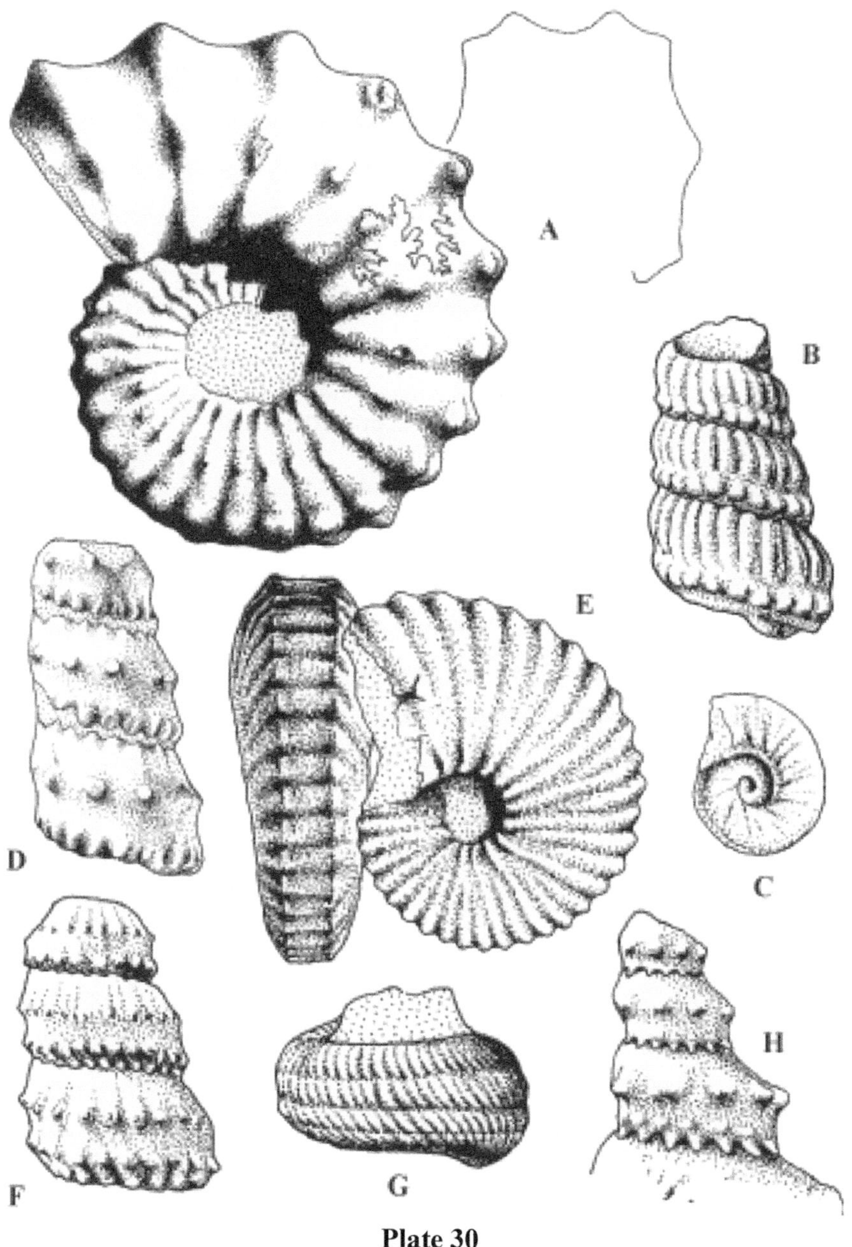

Plate 30

Plate 31

Early - Middle Cenomanian

Ammonites and bivalves from the Mzinene Formation

A. *Gentoniceras paucinodatum* (Crick).
B. *Rhynchostreon suborbiculatum* (Lamarck).
C. *Phyllopachyceras whiteavesi* (Kossmat), x1.5.
D. *Desmoceras latidorsatum* (Michelin), this species ranges into the Albian.
E. *Neithea coquandi* (Zittel).
F. *Trigonarca* cf. *ligeriensis* (d'Orbigny).
G. *Euturrilites scheuchzerianus* (Bosc).
H. *Acanthoceras latum* Crick (syn. *A. flexuosum* Crick).
I. *Forbesiceras sculptum* Crick (syn. *F. nodosum* Crick).

All x1 unless indicated otherwise.

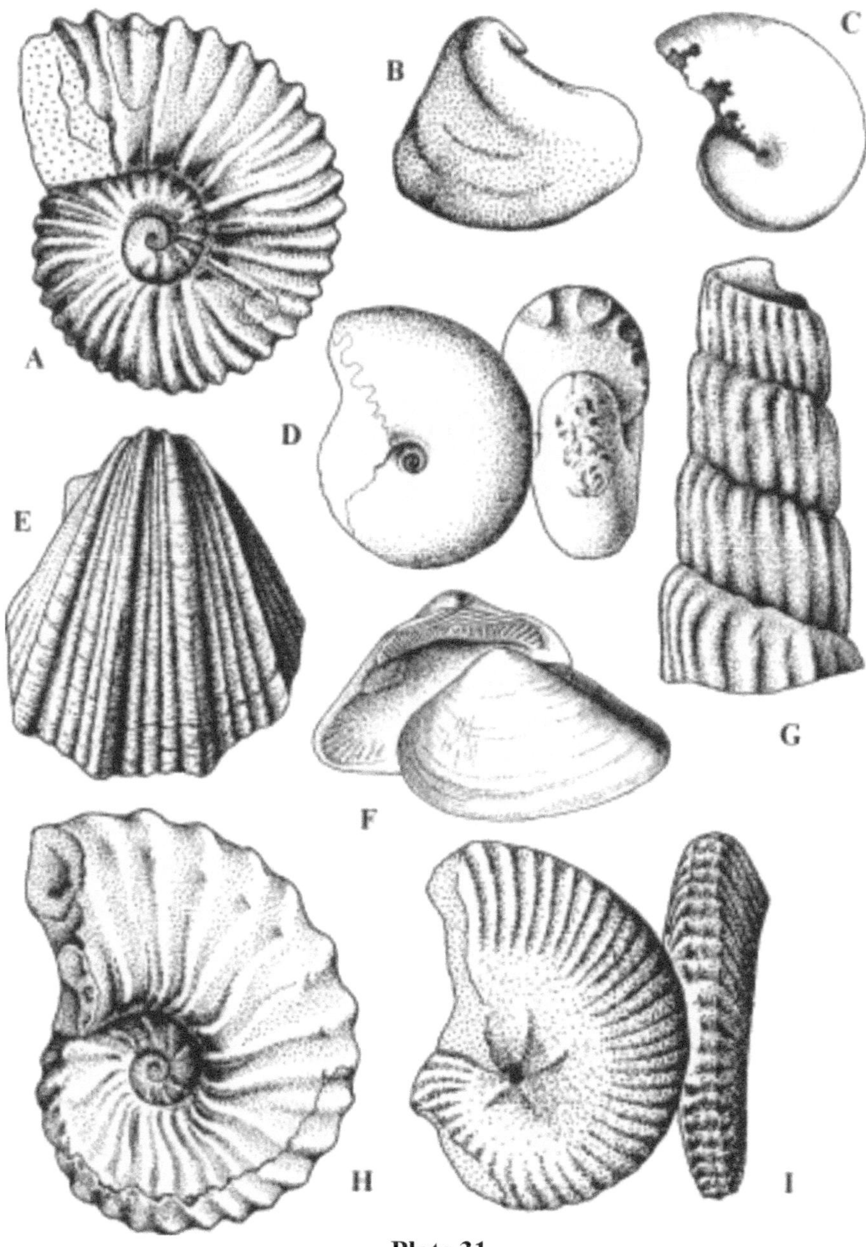

Plate 31

Plate 32

Middle Cenomanian

Ammonites and brachiopod from the Mzinene Formation

A. *Acanthoceras latum* Crick.
B. *Praelongithyris? skoenbergensis* (Muir-Wood).
C. *Anagaudryceras sacya* (Forbes).
D. *Proeucalycoceras choffati* (Kossmat).
E. *Protacanthoceras subwaterloti* (Venzo).
F. *Borrisiakoceras cf. compressum* Reeside, x2.
G. *Acanthoceras cornigerum* Crick.
H. *Acanthoceras* sp. juv.

All x1 unless indicated otherwise.

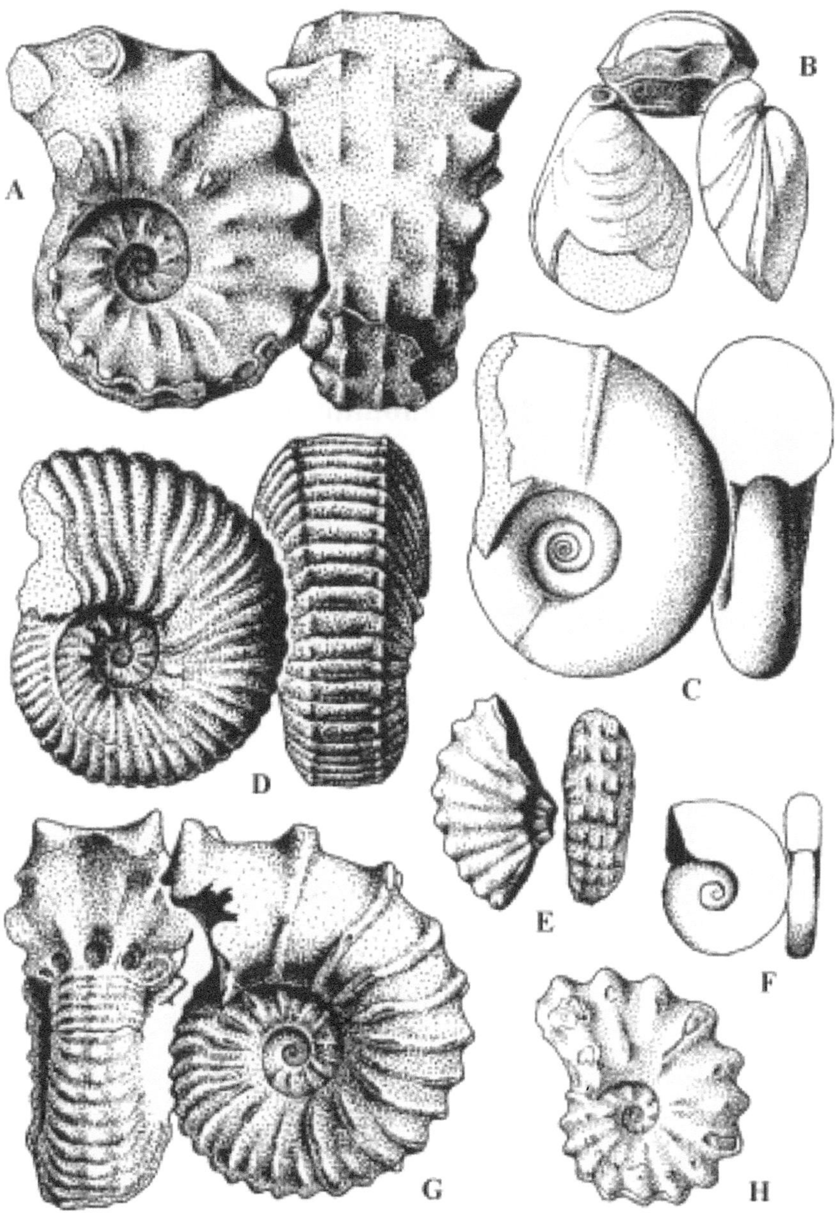

Plate 32

Plate 33

Middle Cenomanian

Ammonites and bivalve from the
Mzinene and Sumbe Formations

A. *Mantelliceras nitidum* (Crick).
B. *Costagyra olisiponensis* (Sharpe).
C. *Hypoturrilites nodiferus* (Crick).
D. *"Knemiceras" cornutum* Crick.
E. *Anisoceras plicatile* (J. Sowerby).
F. *Turrilites acutus* Passy.
G. *Cunningtoniceras meridionale* (Kossmat) (after Cooper), x0.67.

All x1 unless indicated otherwise.

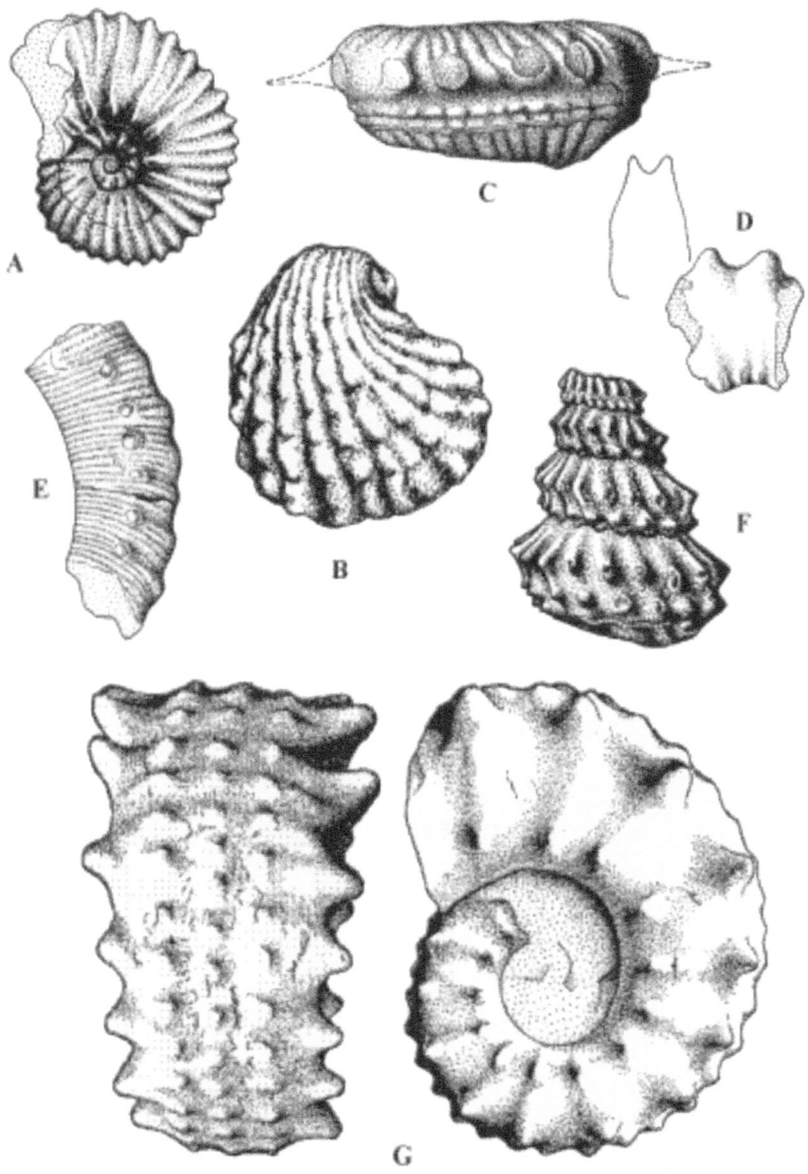

Plate 33

Plate 34

Late Cenomanian

Ammonites and gastropod from the Salinas Formation

A. *Calycoceras naviculare* (Brongniart), x0.75.
B-C. *Kanabiceras septemseriatum* (Cragin), x0.75.
D. *Drepanochilus* sp.
E-F. *Pseudocalycoceras angolense* (Spath).
G. *Metoicoceras geslinianum* (d'Orbigny), x0.8.

All x1 unless stated otherwise.

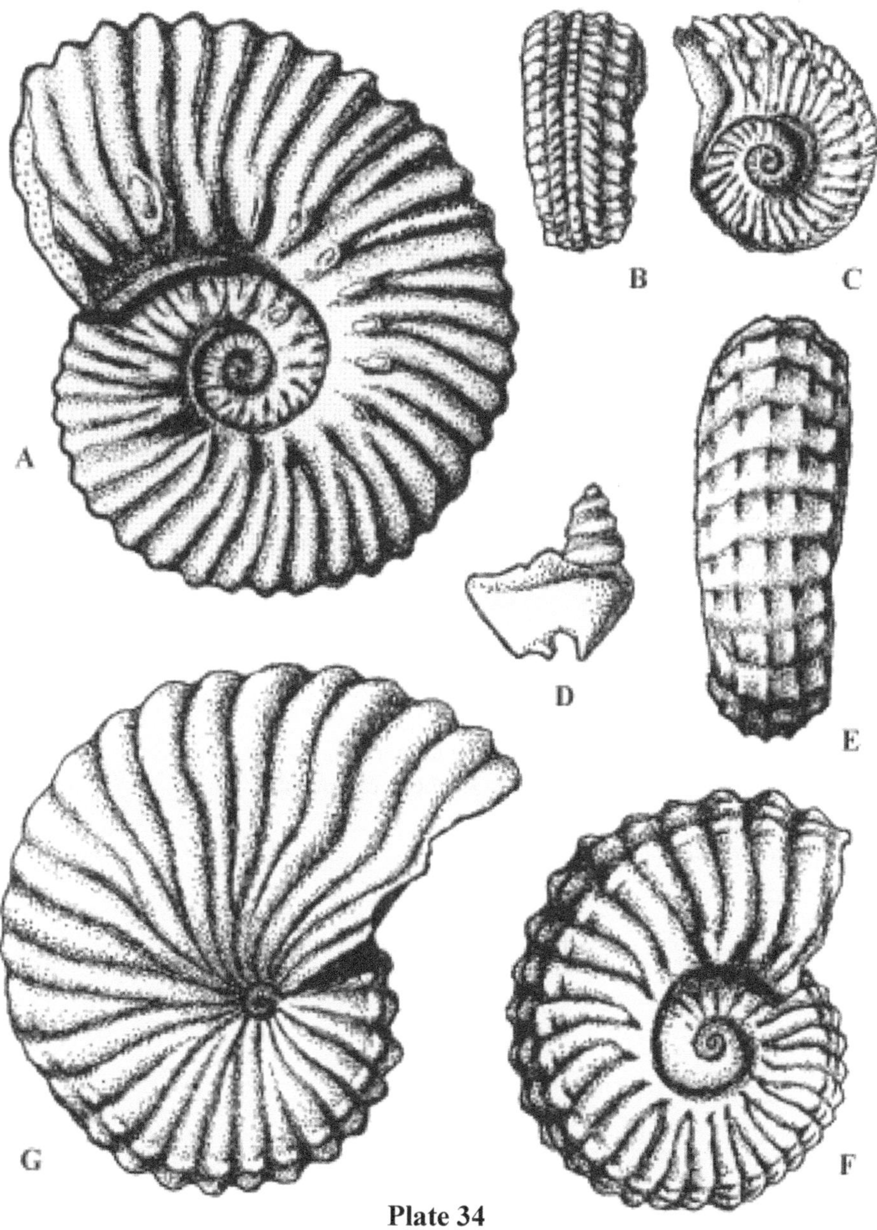

Plate 34

Plate 35

Turonian - Coniacian

Ammonites and bivalve from the
Salinas and Itombe Formations

A-B. *Prionocycloceras carvalhoi* (Howarth), x0.75.
C. *Morrowites mocamedensis* (Howarth), x0.5.
D-E. *Subprionocyclus massoni* (Basse de Menorval).
F-G. *Mossamedites serratocarinatus* (Kennedy & Cobban).
H. *Modiolus typicus* (Forbes).
I-J. *Pseudocucullaea lens* Solger.

All x1 unless stated otherwise.

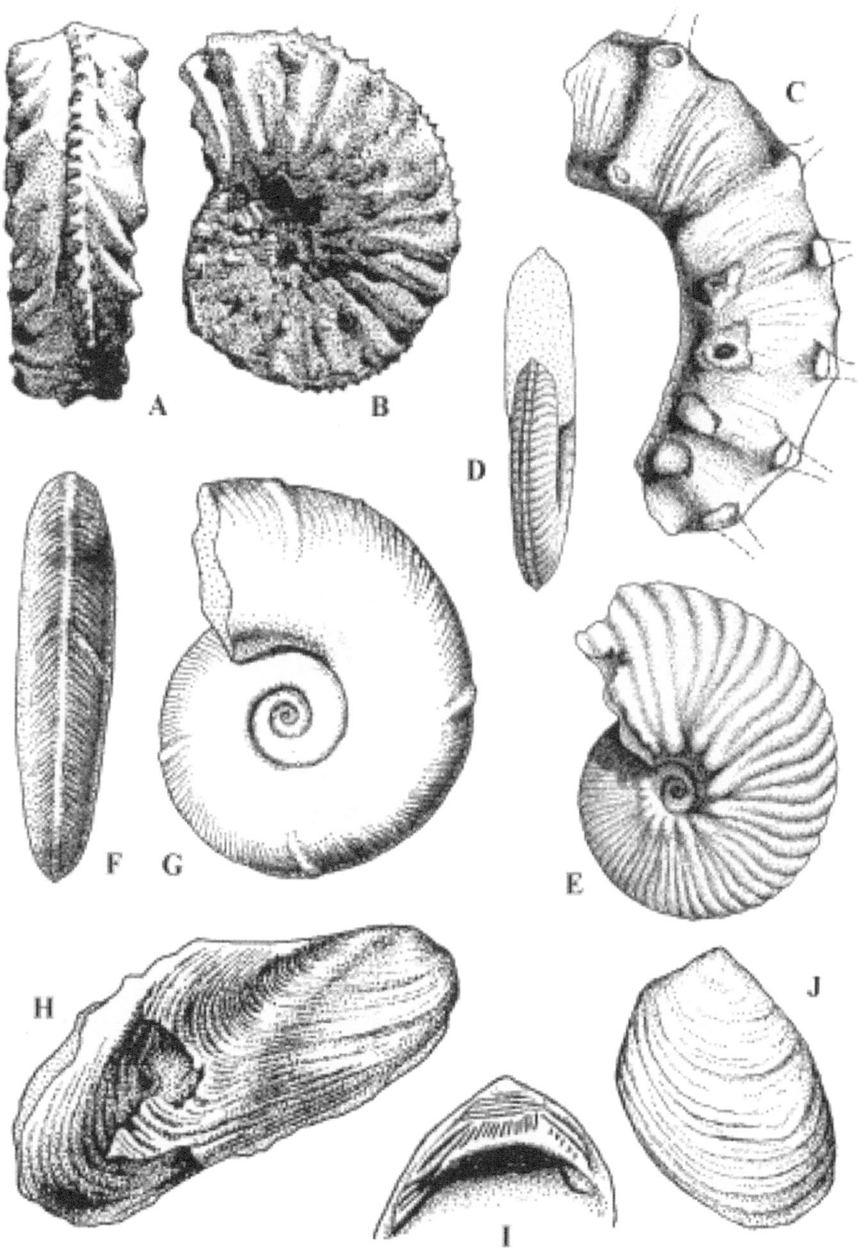

Plate 35

Plate 36

Late Turonian

Mosasaurs, turtle and chondrichthyan teeth
from the Itombe Formation

A. *Angolasaurus bocagei* Antunes, x0.15.
B. *Prognathodon kianda* Schulp, Mateus, Polcyn, Jacobs & Morais, x0.25.
C. *Angolachelys mbaxi* Mateus, Jacobs, Polcyn, Schulp, Neto, Antunes, x0.25.
D. *Ptychodus mortoni* (Mantell), x3.
E. *Paranomotodom angustidens* (Reuss), x2.

All x1 unless stated otherwise.

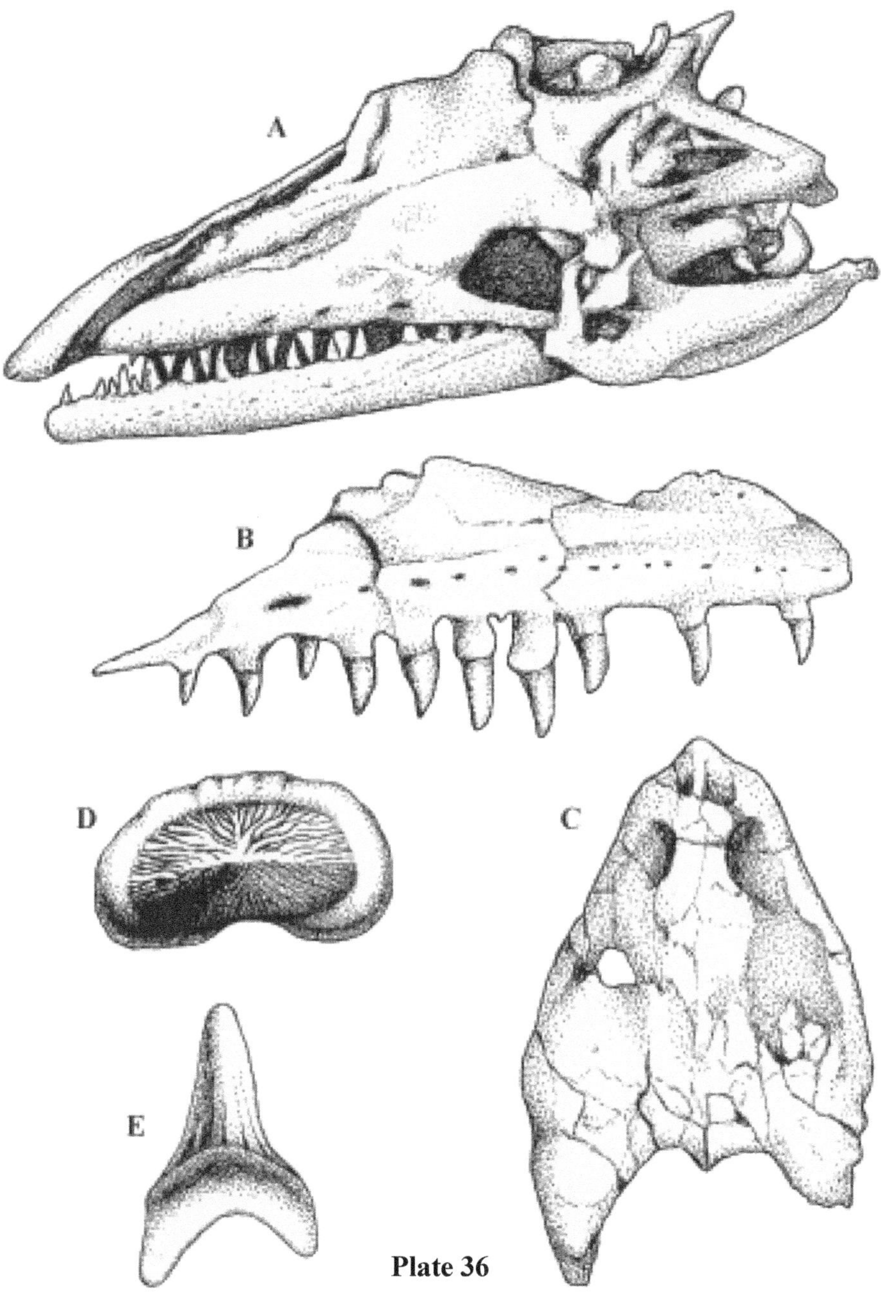

Plate 36

Plate 37

Early - Middle Coniacian

Ammonites and bivalves from the St Lucia
and Salinas Formations

A. *Mesopuzosia indopacifica* (Kossmat).
B. *Kossmaticeras* cf. *theobaldianum* (Stoliczka), close to *K. recurrens* (Kossmat).
C. *Eubostrychoceras indopacificum* Matsumoto.
D. *Proplacenticeras umkwelanense* (Etheridge).
E. *Cremnoceramus waltersdorfensis* (Andert).
F. *Acanthotrigonia shepstonei* (Baily).
G. *Veniella forbesiana* (Stoliczka).

All x1 unless indicated otherwise.

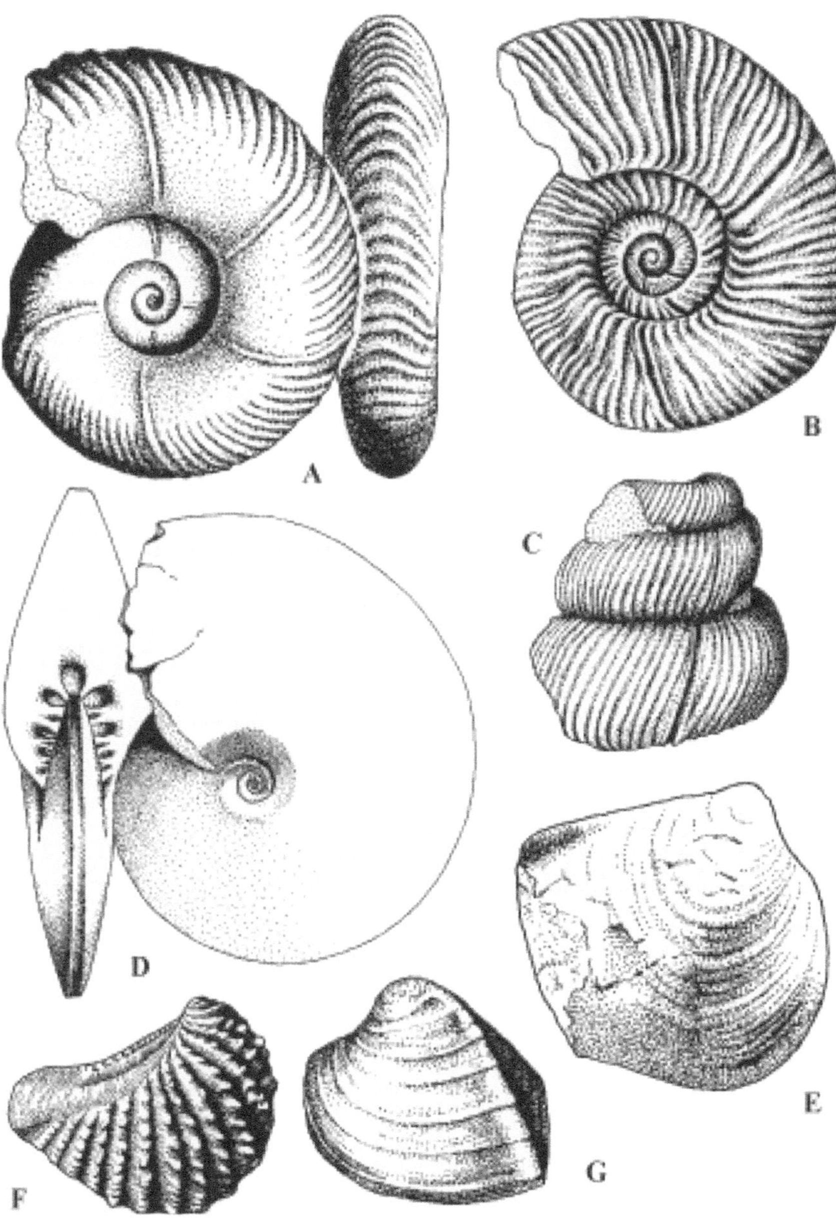

Plate 37

Plate 38

Early - Middle Coniacian

Ammonites and bivalves from the St Lucia Formation

A. *Peroniceras tridorsatum* (Schlueter).
B. *Proplacenticeras kaffrarium (Etheridge)* (syn. *P. subkaffrarium* Spath).
C. *Peroniceras dravidicum* Kossmat.
D. *Tethyoceramus aff. ernsti* (Heinz).
E. *Allocrioceras billinghursti* Klinger.
F. *Inoceramus? pedalinoides* Nagao & Matsumoto.
G. *Forresteria alluaudi* (Boule, Lemoine & Thevenin).

All x1 unless indicated otherwise.

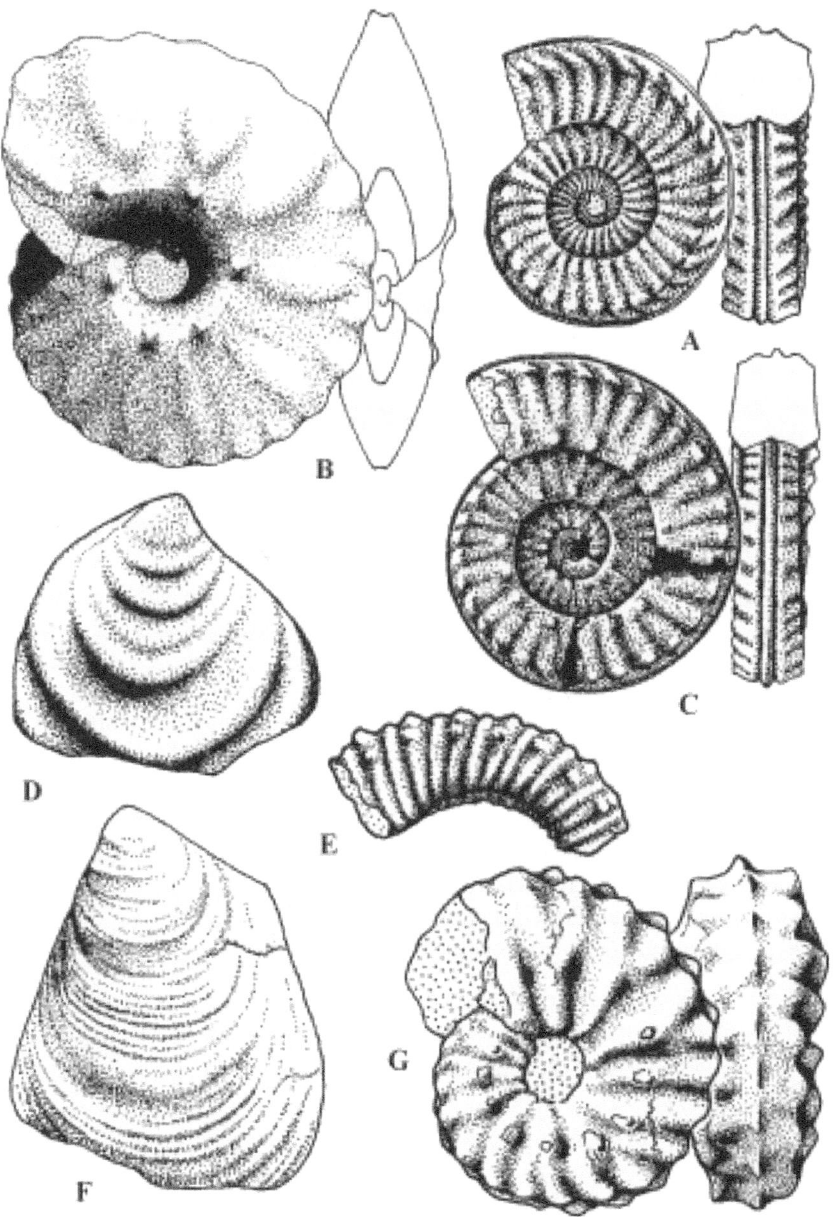

Plate 38

Plate 39

Middle - Late Coniacian

Ammonites and bivalves from the St Lucia Formation

A. *Mytiloides striatoconcentricus* (Guembel).
B. *Neithea quinquecostata* (J. Sowerby); this species ranges into the Campanian.
C. *Itwebeoceras lornae* van Hoepen, x0.75.
D. *Nordenskjoeldia bailyi* (Barnard).
E. *Ceratostreon reticulatum* (Reuss).
F. *Labrostrea umsineniensis* Cooper.
G. *Vultogryphaea goldfussi* (Sobetski), x0.67.
H. *Gauthiericeras obesum* (van Hoepen), x0.5.
I. *Baculites umsinenensis* Venzo, note the oblique dorso-lateral bullae.

All x1 unless indicated otherwise.

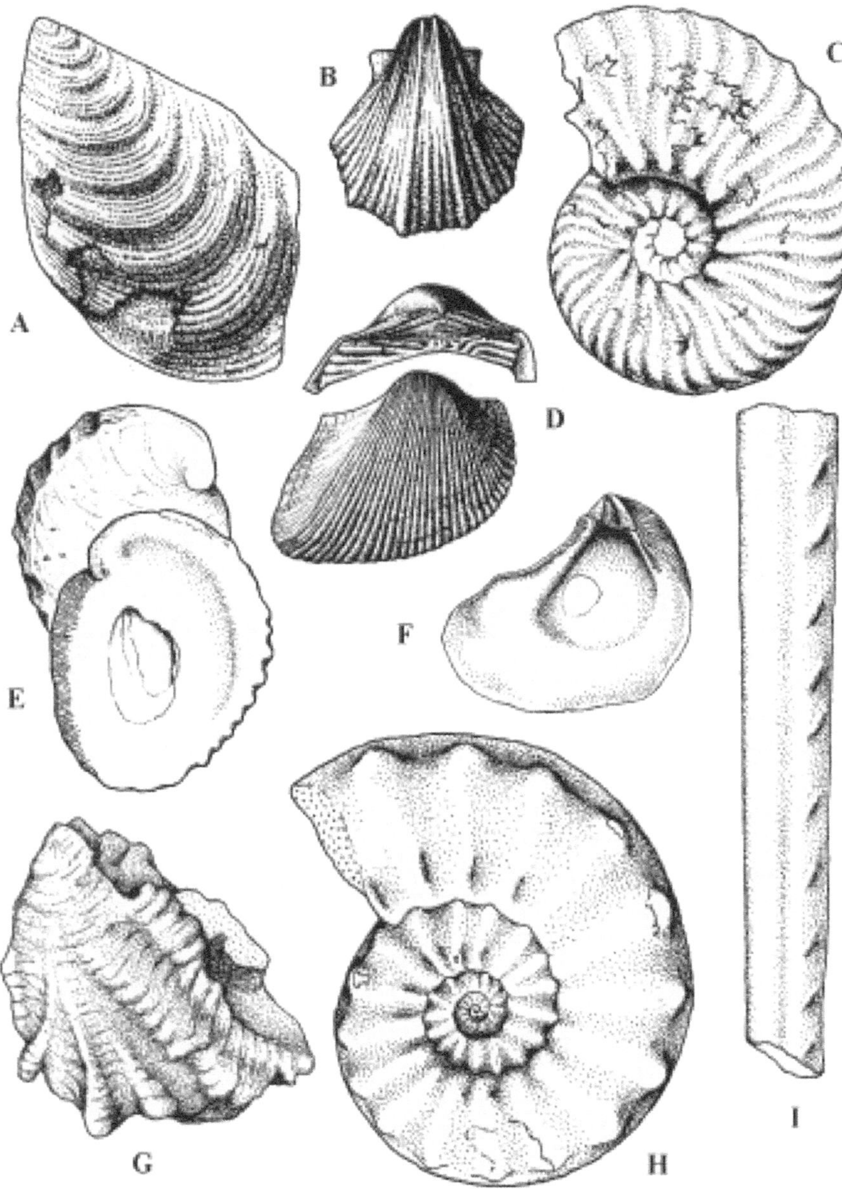

Plate 39

Plate 40

Late Coniacian - Middle Santonian

Ammonites, bivalves, gastropods,echinoid and shark's tooth from the St Lucia, Mzamba and Baba Formations

A. *Texanites vanhoepeni* Klinger & Kennedy, x0.2.
B. *Labrostrea sanctaelucialis* Cooper.
C. *Palaeomoera haughtoni* Rennie.
D. *Hemiaster forbesi* (Baily).
E. *Perissoptera odonnelli* Rennie.
F. *Pseudoschloenbachia umbulazi* (Baily).
G. *Texanites soutoni* (Baily), x0.25.
H. *Cretolamna basalis* (Egerton).
I. *Paleopsephaea odonnelli* Rennie.

All x1 unless indicated otherwise.

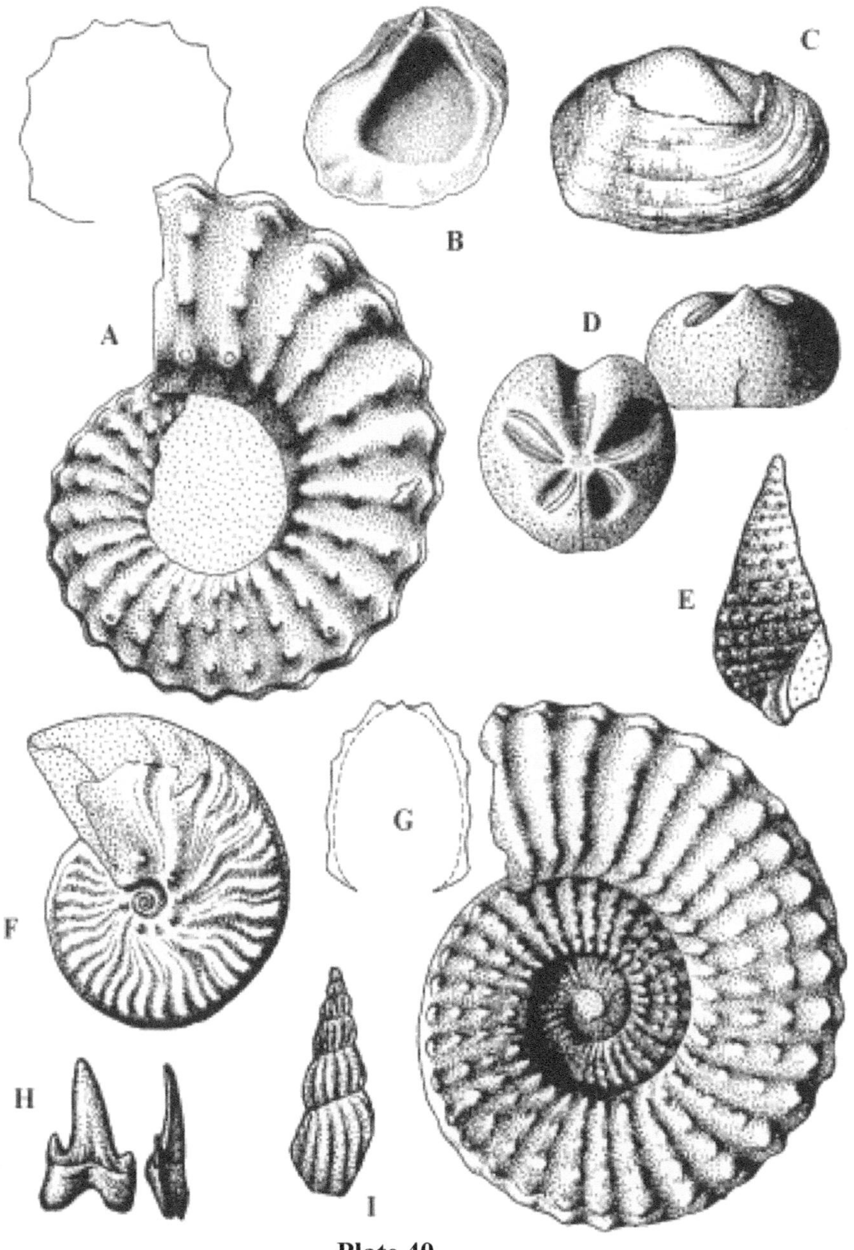

Plate 40

Plate 41

Santonian - Early Campanian

Ammonites and gastropod from the St Lucia,
Mzamba and Baba Formations

A. *Gaudryceras mite* (von Hauer), x0.8; ranges into the Middle Campanian.
B. *Pseudoschloenbachia griesbachi* van Hoepen, typical of the basal Campanian.
C. *Pseudoxybeloceras quadrinodosum* (Jimbo), x0.75.
D. *Baculites capensis* (Baily), common in the Early Santonian.
E. *Gardeniceras gardeni* (Baily), x0.75.
F. *"Pseudomelania" egitoensis* Rennie.

All x1 unless indicated otherwise.

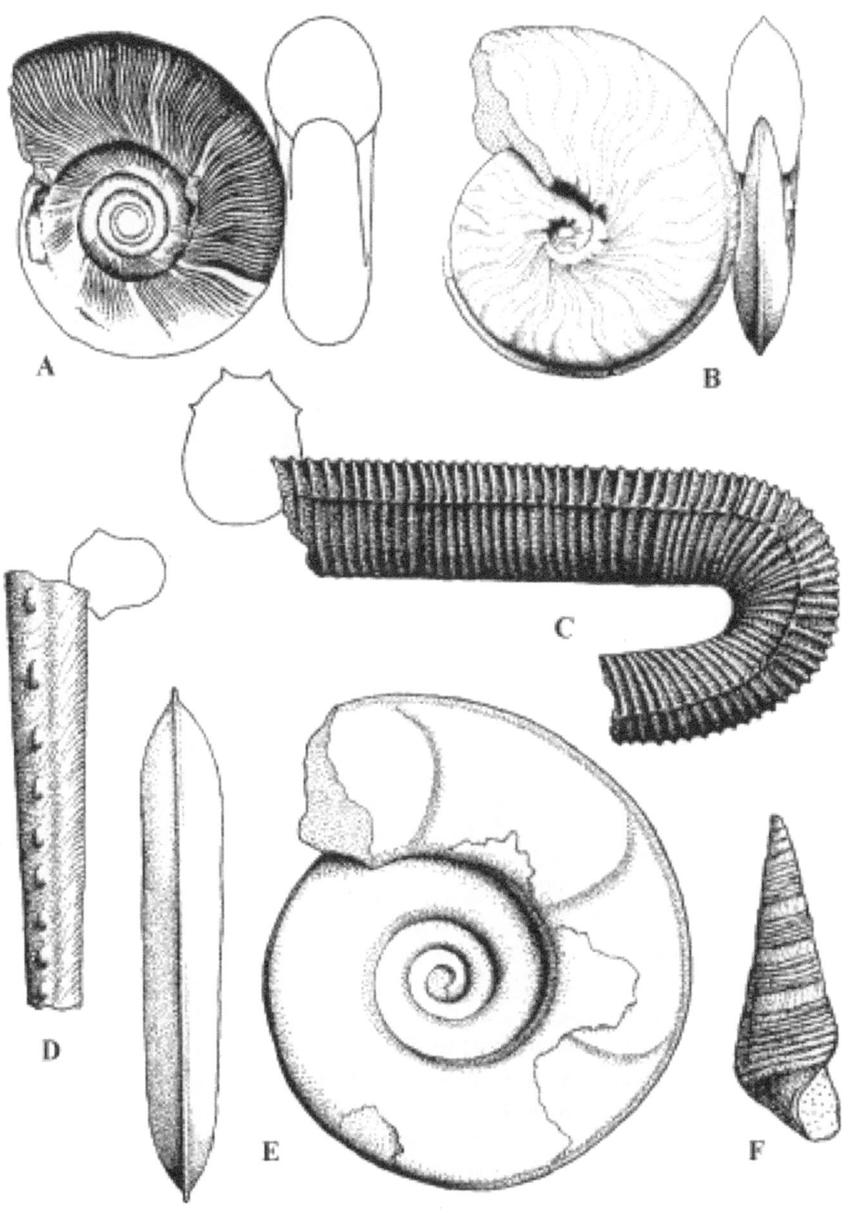

Plate 41

Plate 42

Santonian

Bivalves, ammonites, gastropods and echinoids
from the Mzamba Formation

A. *Hyporbulites woodsi* (van Hoepen), x1.3.
B. *Doyzia lenticularis* (Goldfuss), x1.5.
C. *Acanthocardia denticulata* (Baily).
D. *Trachycardium griesbachi* (Woods).
E. *Texasia cricki* (Spath).
F. *Lupira? meridionalis* (Woods), x1.5.
G. *Astarte griesbachi* Woods, x2.
H. *Cassidulus umbonatus* Woods.
I. *Perissoptera bailyi* (Etheridge).
J. *Natalites africanus* (van Hoepen).
K. *Cardiaster africanus* Woods, x1.5.
L. *Macoma? papyracea* Rennie.
M. *Protopirula capensis* (Rennie) - a juvenile cypraeid.
N. *Pleuromya renniei* da Silva.
O. *Paleopsephaea scalaris* Rennie.

All x1 unless indicated otherwise.

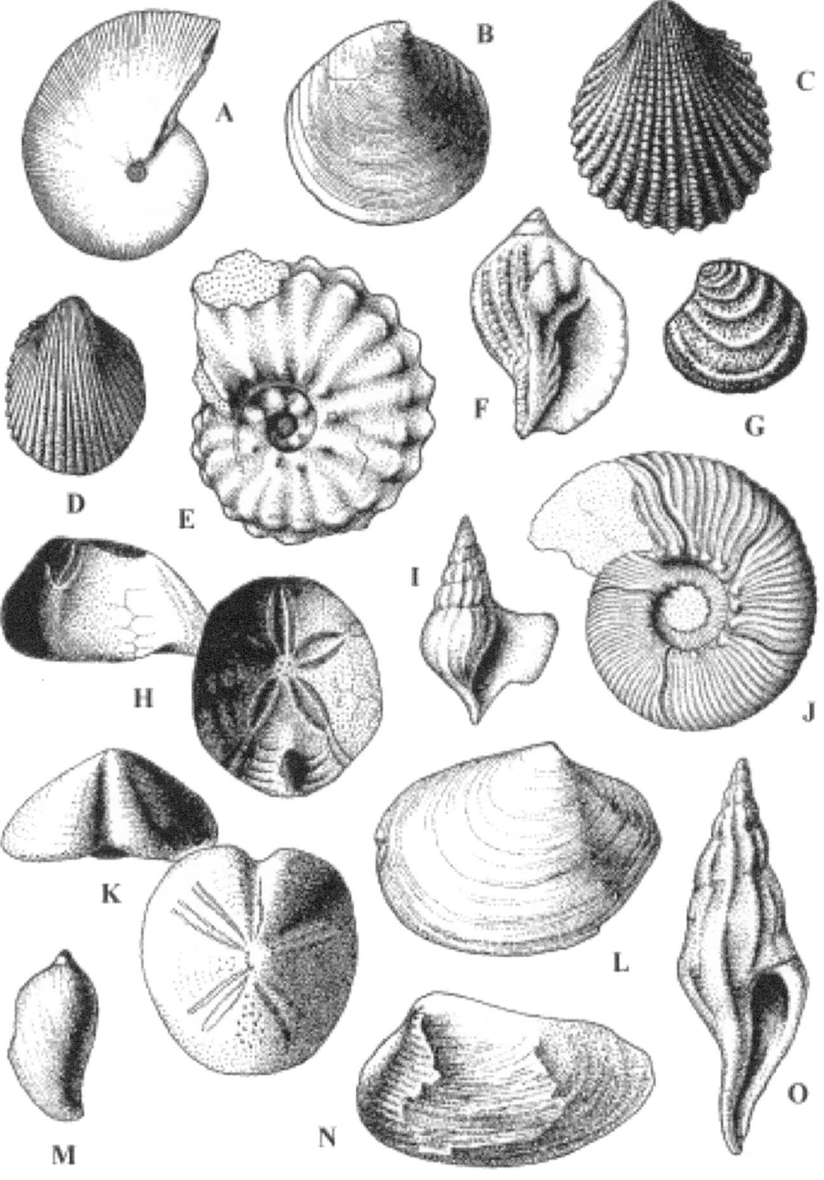

Plate 42

Plate 43

Santonian - Early Campanian

Bivalves, gastropods and shark's tooth from the Mzamba Formation

A. *Cladoceramus undulatoplicatus* (Roemer), x0.3.
B. *Dosiniopsis geversi* Rennie.
C. *Cretoxyrhina* aff. *mantelli* (Agassiz).
D. *Goniomya umzambiensis* Rennie.
E. *Ringicula woodsi* Rennie, x3.
F. *Agathodonta africana* (Rennie).
G. *Pyropsis africana* Woods.
H. *Zaria bonei* (Baily).
I. *Afrocypraea chubbi* (Rennie).
J. *Aphrodina euglypha* (Woods).
K. *Anisodonta umzambiensis* Rennie.
L. *Cercomya arcuata* (Forbes).
M. *Trigonarca capensis* (Griesbach).

All x1 unless indicated otherwise.

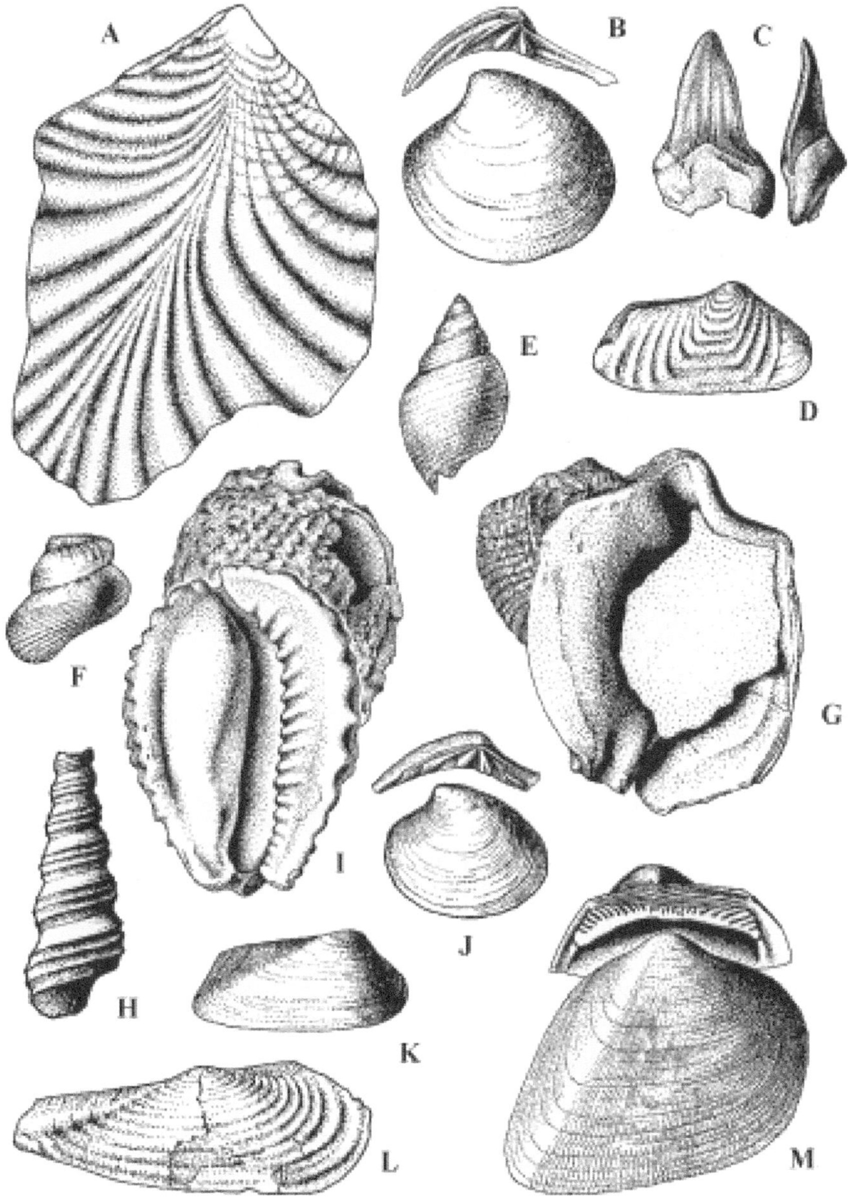

Plate 43

Plate 44

Santonian - Early Campanian

Bivalves, gastropods and ammonite
from the Mzamba and Baba Formations

A. *Spondylus* cf. *calcaratus* Forbes.
B. *Camptonectes kaffraria* Rennie.
C. *Patellona kaffraria* (Rennie).
D. *Protocardia umkwelanensis* Etheridge.
E. *Liopeplum capensis* (Woods).
F. *Gymnarus auriculatus* (Woods).
G. *Trachycardium reynoldsi* Rennie.
H. *Rastellum deshayesi* (Fischer de Waldheim), x0.75.
I. *Linotrigonia itongazi* (de Little).
J. *Deussenia rigida* (Baily).
K. *Hemifusus umzambiensis* (Rennie).
L. *Ambigostrea* cf. *arcotensis* (Stoliczka).
M. *Damesites compactus* (van Hoepen).
N. *Acutostrea* cf. *bucheroni* (Coquand).

All x1 unless indicated otherwise.

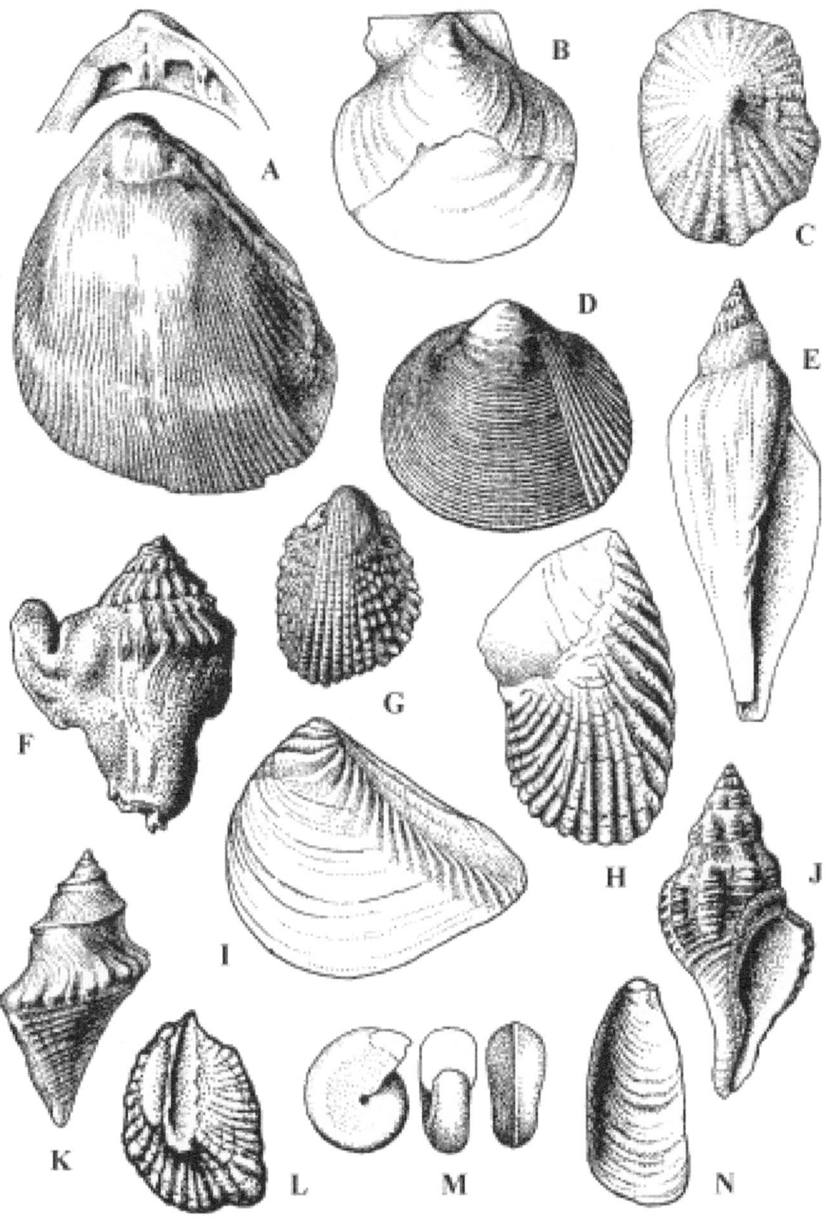

Plate 44

Plate 45

Santonian - Middle Campanian

Ammonite, gastropods, bivalves and echinoids
from the Mzamba and Quimbala Formations

A. *Pseudophyllites indra* (Forbes), x0.4.
B. *Paosia dutoiti* (Rennie).
C. *Pseudomelania sutherlandi* (Baily).
D. *Austromactra? rogersi* (Rennie).
E. *Trochactaeon woodsi* (Rennie).
F. *Roundairia drui* (Munier-Chalmas).
G. *Ampullina? multistriata* (Baily).
H. *Trigonarca elongata* Rennie.
I. *Crassatellites africanus* (Woods).
J. *Tholaster carvalhoi* Greyling & Cooper, x1.5.
K *Leiostomaster angolanus* Greyling & Cooper, x1.5.

All x1 unless indicated otherwise.

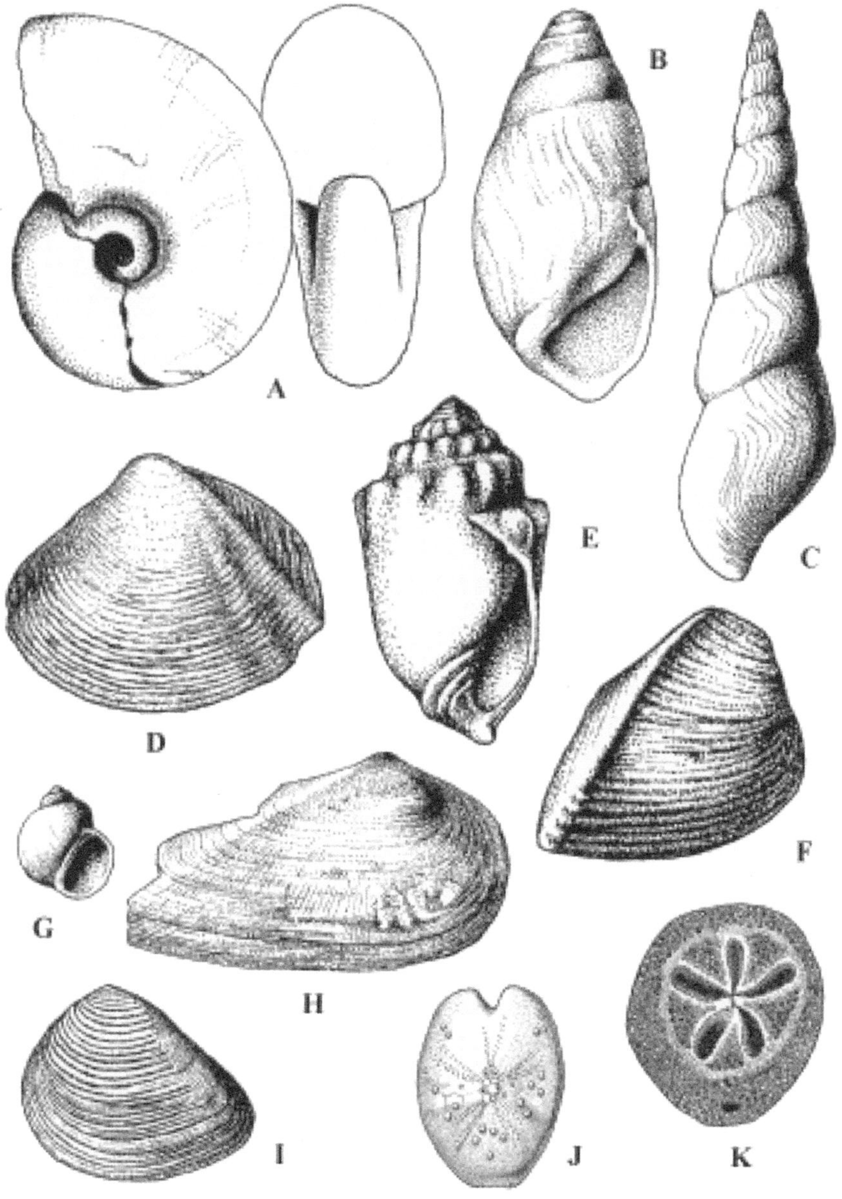

Plate 45

Plate 46

Campanian

Ammonites, bivalves and brachiopod from the
Mzamba, St Lucia and Rio Dande Formations

A. *Nostoceras rotundatum* Howarth.
B. *Zulostrea zulu* Cooper.
C. *Exogyra ponderosa nibelaensis* Cooper.
D. *Eulophoceras natalense* Hyatt.
E. *Didymoceras depressum sanctaeluciensis* Klinger.
F. *Linotrigonia nibelaensis* Cooper.
G. *Cyrtothyris undisona* Cooper.

All x1 unless indicated otherwise.

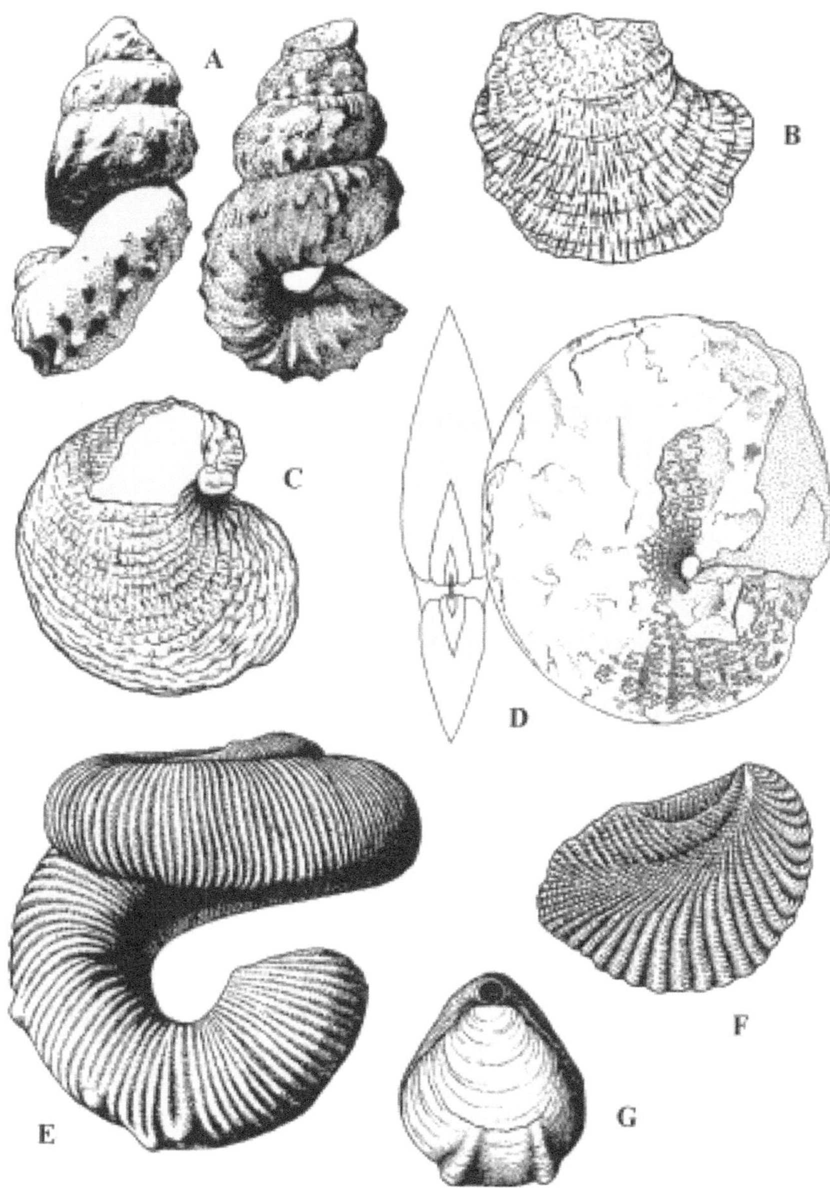

Plate 46

Plate 47

Early - Middle Campanian

Oyster and ammonites from the St Lucia Formation

A. *Gryphaeostrea canaliculata* (J. de C. Sowerby).
B. *Submortoniceras woodsi* Spath, x0.67.
C. *Baculites vanhoepeni* Venzo.
D. *Diaziceras tissotiforme* Spath.

All x1 unless indicated otherwise.

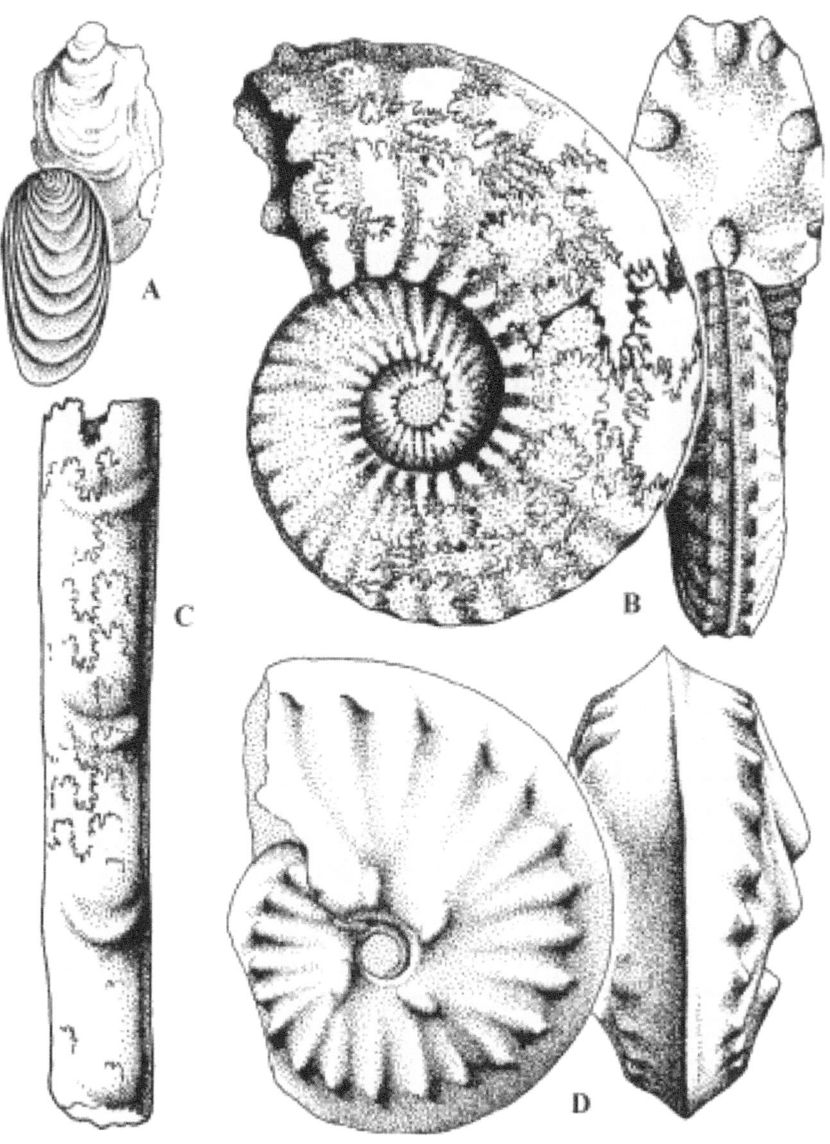

Plate 47

Plate 48

Middle Campanian - Early Maastrichtian

Ammonites and bivalve
from the St Lucia, Quimbala and Rio Dande Formations

A. *Manambolites dandensis* Howarth, x0.5.
B. *Cirroceras depressum* (Howarth).
C. *Didymoceras subtuberculatum* Howarth.
D. *Neithea striatocostata* (Goldfuss), x3.
E. *Phylloptychoceras dandense* Howarth.
F. *Menuites macgowani* Haughton, x0.5.
G. *Kitchinites angolensis* (Spath) (syn. *Oiophyllites angolensis* Spath, *K. angolensis* Howarth).

All x1 unless indicated otherwise.

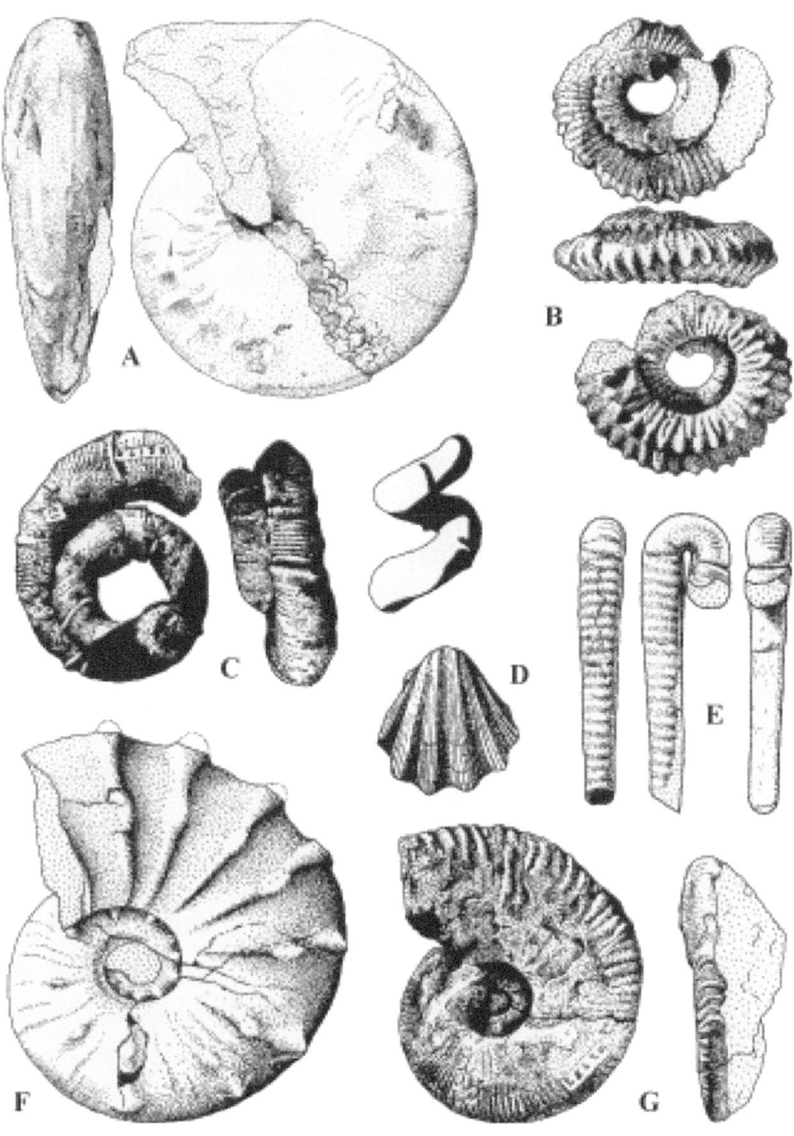

Plate 48

Plate 49

Campanian-Maastrichtian

Ammonites and echinoid from the
St Lucia and Incomanine Formations

A. *Bolbaster madagascariensis* (Cottreau), x1.5.
B. *Vepricardium? kossmati* (Schlösser), x0.8.
C. *Anapachydiscus subdulmensis* (Venzo).
D. *Glycymeris merenskyi* (Schlösser), x0.8.
E. *Desmophyllites larteti* (Seunes).
F. *Eubaculites carinatus* (Morton), x0.8.

All x1 unless indicated otherwise.

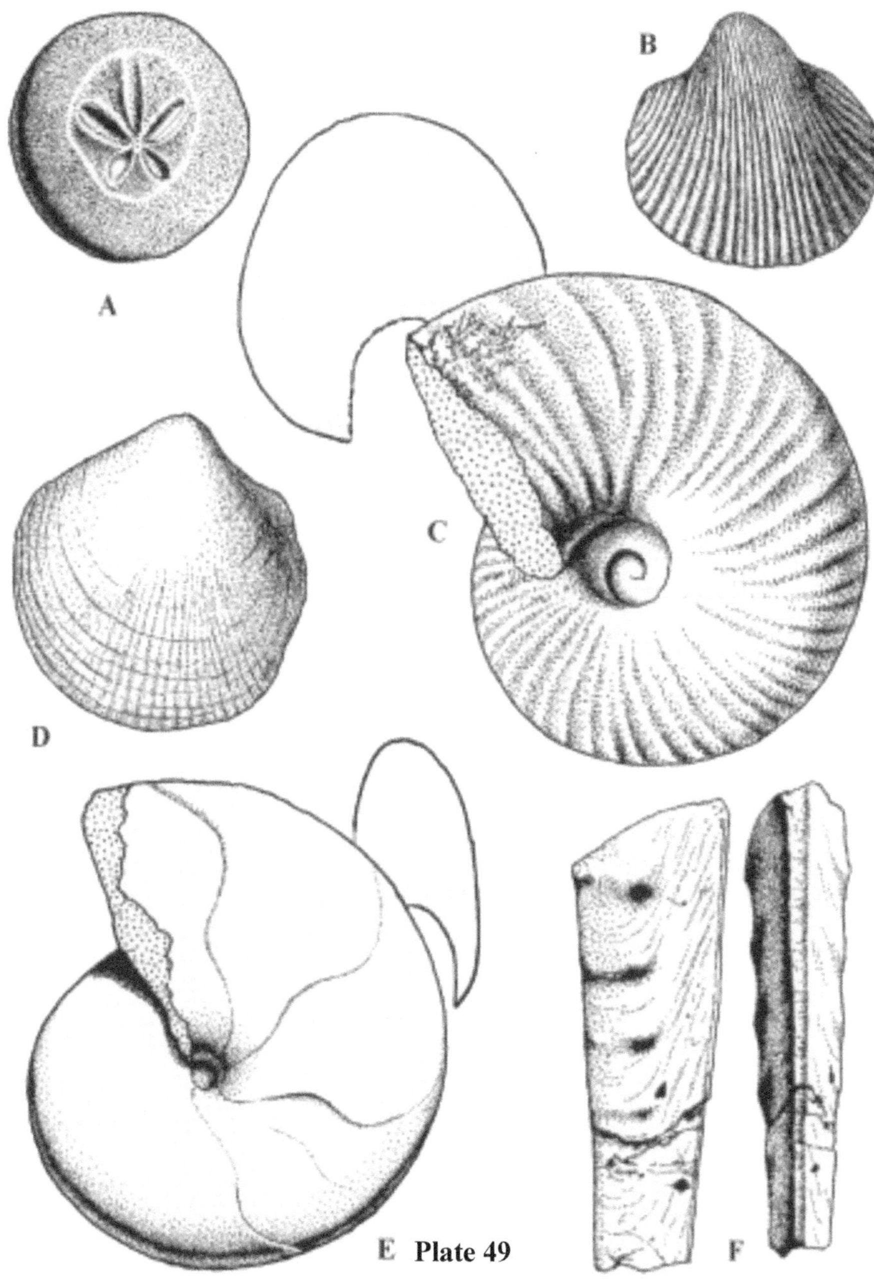

Plate 49

Plate 50

Late Campanian-Maastrichtian

Ammonites, coral and shark's tooth from the
St Lucia, Incomanine, Rio Dande and Mocuio Formations

A. *Pachydiscus australis* Henderson & McNamara, x0.5.
B. *Solenoceras binodosum* Haughton.
C. *Neophylloceras ultimatum* Spath.
D. *Axonoceras tenue* Haas.
E. *Saghalinites cala* (Forbes).
F. *Ceratotrochus mennelli* Gregory, x2.
G. *Squalicorax pristodontus* (Agassiz), x1.5.

All x1 unless indicated otherwise.

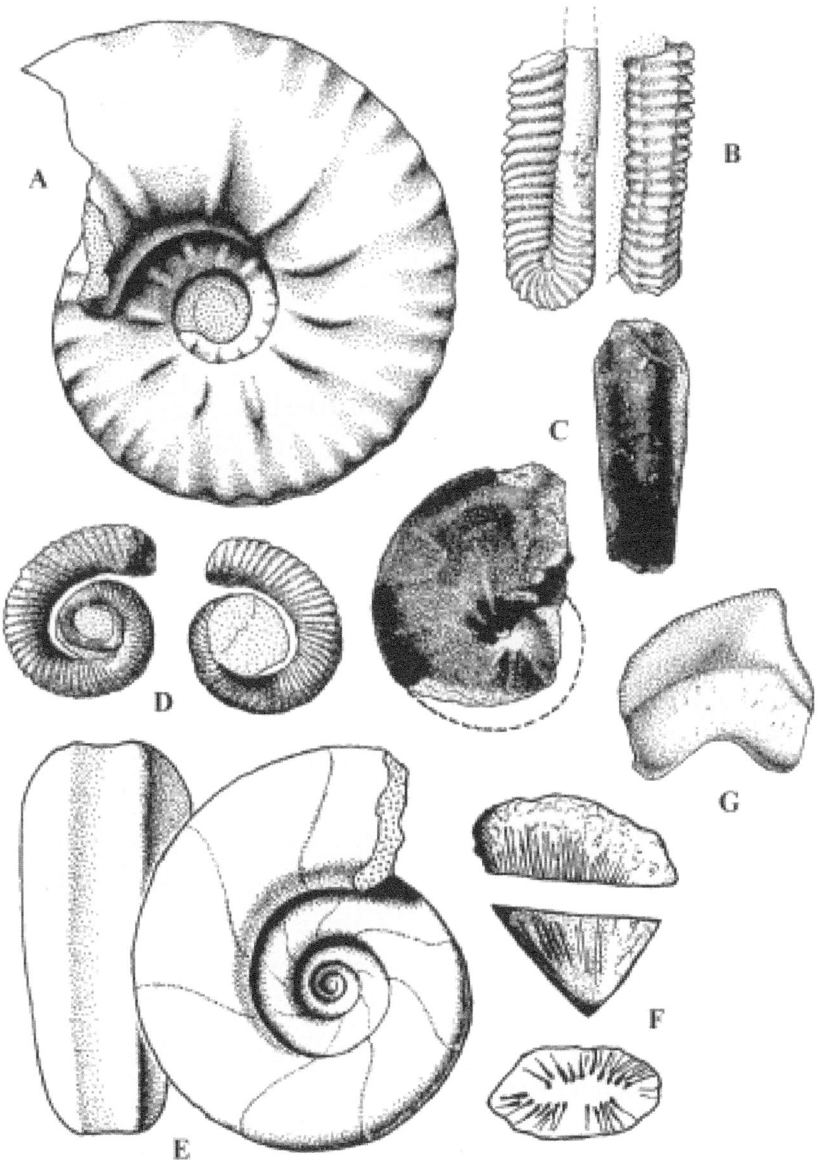

Plate 50

Plate 51

Maastrichtian fossils

Nautiloids, ammonites and bivalves
from the Grudja, Incomanine and St Lucia Formations

A. *Hercoglossa sheringomensis* Crick.
B. *Phygraea vesicularis zululandensis* Cooper.
C. *Hercoglossa mazambensis* Crick.
D. *Eubaculites carinatus* (Morton).
E. *Epiphylloceras surya* (Forbes).
F. *Trochoceramus andimakenensis* (Sornay).

All x1 unless indicated otherwise.

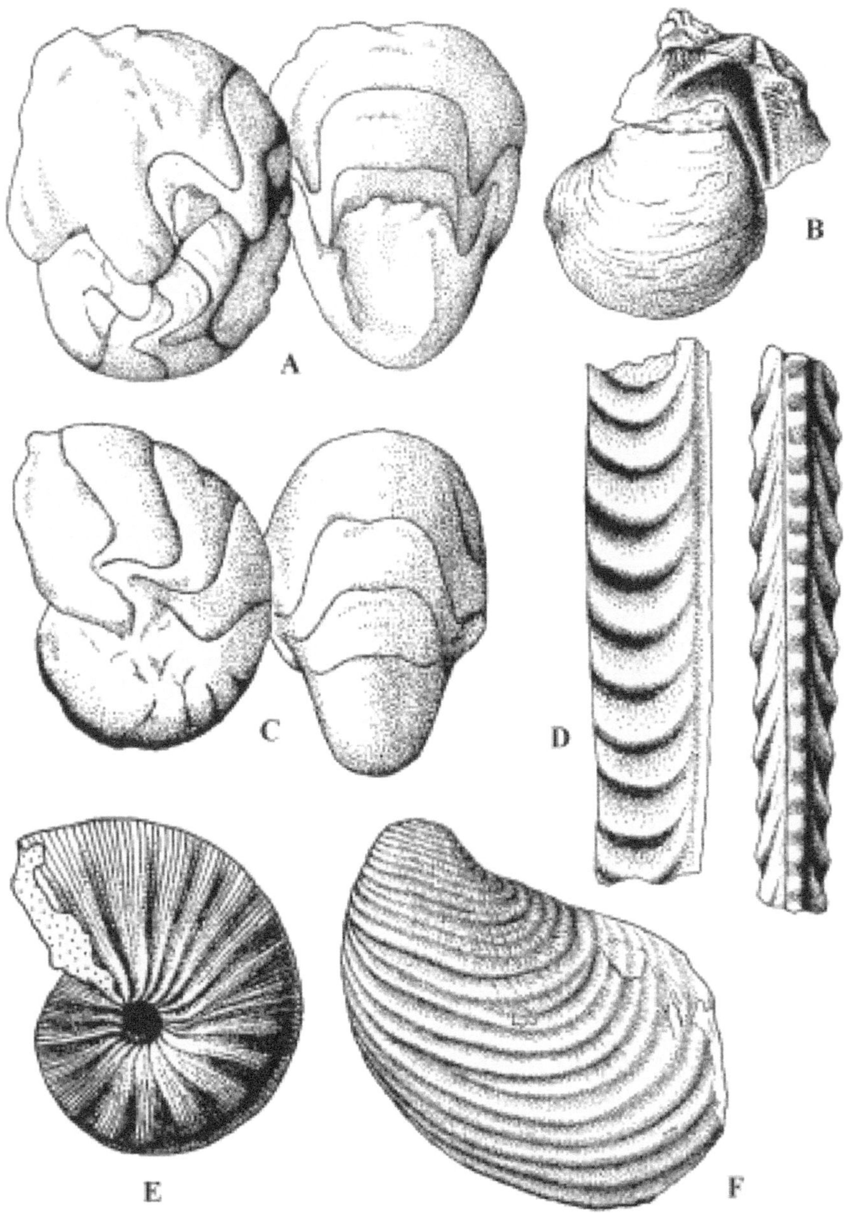

Plate 51

Plate 52

Maastrichtian

Bivalves and gastropods from the Incomanine
and St Lucia Formations

A. *Trigonocallista woodsi* Rennie.
B. *Venericardia* cf. *beaumonti* (d'Archaic & Haime).
C. *Euspira tealei* (Cox), x2.
D. *Lycettia* cf. *arrialoorensis* (Stoliczka), x2.
E. *Tudicla krenkeli* Cox, x2.
F. *Haustator scala* (Cox).
G. *Theodoxus? choffati* (Cox).
H. *Cerithidea? haughtoni* (Cox), x1.5.
I. *Gryphaeostrea canaliculata* (J. Sowerby).
J. *Pollia? gregoryi* (Cox), x5.
K. *Nuculana* cf. *scaphuloides* (Stoliczka), x2.
L. *"Cardium" haughtoni* Rennie.
M. *Nuculana andradei* Rennie, x4.
N. *Camptonectes curvatus* (Geinitz).
O. *Littorinopsis africa* Cox.
P. *Velostreon umfolozianum* Cooper.
Q. *Incomatiella incomatensis* (Rennie), x0.75.
R. *Linearia incomatensis* Rennie.
S. *Agerostrea rouxi* (Douville).

All x1 unless indicated otherwise.

Plate 52

Index of lithostratigraphy

Addo Member ... 6
Amsterdamhoek Member ... 6
Arnot Pipe ... 4
Baba Formation ... 24,25
Bentiaba Group ... 24
Bero Conglomerate ... 25
Bezuidenhouts Member ... 5
Binga Formation ... 27
Boane Formation ... 21
Brenton Formation ... 8
Bumbeni Complex ... 11
Cabo Ledo Formation ... 27
Cacoba Formation ... 28
Cape Peninsula dyke swarm ... 4
Colchester Member ... 5
Calonda Formation ... 29
Cangulo Formation ... 24
Catoca kimberlite ... 29
Catumbela Member ... 27
Chela Sandstone ... 27
Chidziwa Formation ... 23
Chiswiti Formation ... 23
Continental Intercalaire ... 29
Coporolo Formation ... 26
Cuio Formation ... 26
Cuvo Formation ... 27
Dasdab Formation ... 3
Dombe Formation ... 26
Domo Formation ... 22
Dondo Formation ... 27
Dinosaur Beds, Malawi ... 29
Enon Formation ... 5
Etendeka Formation ... 3
Fenda Formation ... 11
Flamingo Formation ... 27
Formação com *Nerinea* ... 26
Formação com *Pholadomya* ... 26
Galula Formation ... 30
Golweni Formation ... 11
Giraul Conglomerate ... 25
Grootberg Formation ... 3
Group I kimberlites ... 4
Group II kimberlites ... 4
Grudja Formation ... 22
Hluhluwe Group ... 11,12
Igoda Formation ... 8
Inamagando Beds ... 25
Incomanine Formation ... 22
Infanta Shale ... 6
Inhandue Formation ... 19
Itombe Formation ... 28
Kalahari Group ... 4,24
Karonga Formation ... 29
Kirkwood Formation ... 5
Kikundi Formation ... 30
Kiturika Formation ... 30
Koako Group ... 3
Kuleni Rhyolite ... 11
Kwango Group ... 29
Kwanza Group ... 26
Loeme Salt Formation ... 27
Longa Formation ... 27
Lourenço Marques Conglomerate ... 22
Lower Need's Camp Formation ... 8
Lupata Group ... 19
Maculungo Formation ... 27
Makhatini Formation ... 12
Makonde Formation ... 30
Malvernia Formation ... 24
Maputaland Group ... 11
Maputo Formation ... 20
Mbotyi Formation ... 9
Melkfontein CarbonatiteTuff ... 10
Mocuio Formation ... 24,25,26
Mocungo Member ... 24
Mkuze Formation ... 12
Mngazana Formation ... 8
Mothae kimberlite ... 5
Movene basalts ... 11
Mpilo Formation ... 11
Msunduze Formation ... 11
Munywane Formation ... 11
Mzamba Formation ... 9,10
Mzinene Formation ... 12,13
Namitambo Formation ... 29

N'Golome Shale.. 28
Novo Redondo Formation.......................... 28
Ntandi Formation...................................... 30
Nxwala Rhyolite 11
Ombe Formation 24,25
Oyster shell bed.. 4
Pambala Formation 28
Piambo Formation.................................... 24
Ponta Grossa Member............................... 25
Ponta Negra Member 24
Quiango Formation 27
Quimbala Formation 28
Quissonde Member 27
Red Sandstone Group, Tanzania 30
Richards Bay Formation 12
Ring Syenite..11
Rio Dande Formation................................ 29
Riverview Formation 15
Robberg Formation 7
Rutitrigonia schwarzi Beds....................... 30
São Nicolau Group.................................... 24
Salinas Formation 24,25
Sarula Formation....................................... 19
Sena Sandstone Formation........................ 20
Sena Subgroup 19
Soetgenoeg Member 6
Sorodzi Formation 19
St Lucia Formation 12,16
Sumbe Formation...................................... 28
Sundays River Formation 6
Swartkops Member..................................... 5
Tambara Formation................................... 19
Teba Formation 29
Tendaguru Supergroup.............................. 29
Tuenza Formation 27
Uitenhage Group... 5
Umbilo Formation..................................... 10
Vetmaak Member.. 6
Wanderveld IV Formation........................... 4
Zululand Supergroup.................................11

Index of illustrations

The number in bold refers to the plate, and the letter in normal type to the figure.

A

Acanthocardia denticulata **42**,C
Acanthoceras cornigerum **32**,G
—— *flexuosum* **31**,H
—— *latum* **31**,H,**32**,A
—— sp. juv. **32**,H
Acanthohoplites aschiltaensis **18**,E
—— *bigoureti* **17**,D
Acanthotrigonia shepstonei **37**,F
Acesta obliquissima **7**,E
Achilleoceras erasmusi **27**,A-B
Aconeceras nisus **17**,C
Acrioceras zulu **13**,D
Actaeonella anchietai **29**,A
Actaeonina atherstoni **7**,D
Actinoceramus concentricus **24**,B
—— *salomoni* **19**,D
—— *sulcatus* **22**,F
—— *volviumbonatus* **26**,B
Acutostrea cf. *bucheroni* **44**,N
Aetostreon imbricatum **3**,E
—— *latissimum* **12**,E
Afrocypraea chubbi **43**,I
Agathodonta africana **43**,F
Agerostrea rouxi **52**,S
Allocrioceras billinghursti **38**,E
Allomactra sp. nov. **22**,M
Alopecoceras ankeritteri **20**,E
Ambigostrea cf. *arcotensis* **44**,L
Amphidonte malchusi **26**,C
—— *arduennensis austroafricanum* **21**,E
Anagaudryceras pulchrum **24**,H
—— *sacya* **32**,C
Anapachydiscus subdulmensis **49**,C
Angolachelys mbaxi **36**,C
Angolasaurus bocagei **36**,A
Anisoceras perarmatum **25**,F
—— *plicatile* **33**,E
Anisodonta umzambiensis **43**,K
Angolaites simplex **28**,H
Angolasaurus bocagei **36**,A
Anthonya lineata **7**,G
Aphrodina euglypha **43**,J
Araucarites rogersi **1**,B
Arcomya baini **5**,B
Arctica rugulosa **4**,J
Arestoceras splendidum **23**,A
Audouliceras cf. *ajax* **15**,C
Astarte griesbachi **42**,G
Austromactra? rogersi **45**,D
Austromyophorella dooleyi **11**,B
—— *oosthuizeni* **7**,F
Axonoceras tenue **50**,D

B

Baculites capensis **41**,D
—— *umsinenensis* **39**,I
—— *vanhoepeni* **47**,C
Bathrotomaria uitenhagensis **9**,F
Belemnopsis gladiator **1**,P
Bhimaites spathi **26**,G
Bochianites africanus **6**,C
Bolbaster madagascariensis **49**,A
Borrisiakoceras cf. *compressum* **32**,F
Brachyphyllym sp. **1**,L

C

Cainoceras strigosum **23**,F
Calva cf. *subrotunda* **17**,F
Calycoceras naviculare **34**,A
Camptonectes cottaldinus **6**,H
—— *curvatus* **52**,N

—— *kaffraria* **44**,B
Cancellaria meridionalis **42**,F
Cardiaster africanus **42**,K
"Cardium" haughtoni **52**,L
Cassidulus umbonatus **42**,H
Ceratostreon reticulatum **39**,E
Ceratotrochus mennelli **50**,F
Cercomya arcuata **43**,L
Cerithidea? haughtoni **52**,H
Chalalabelus renniei **13**,F
Cheloniceras gottschei **15**,D
Chlamys? cf. *subacutus* **4**,F
Cirroceras depressum **48**,B
Cladoceramus undulatoplicatus **43**,A
Cladophlebis browniana **1**,N
—— *denticulata atherstoni* **1**,H
Coelastarte coxi **12**,A
Colchidites vulanensis australis **13**,A
Confusiscala borgesi **11**,C
—— *chalalensis* **12**,D
Conites sp. **1**,C
Costagyra olisiponensis **33**,B
Crassatellites africanus **45**,I
Cremnoceramus waltersdorfensis **37**,E
Cretolamna basalis **40**,H
Cretoxyrhina aff. *mantelli* **43**,C
Cryptocrioceras yrigoyeni **12**,B
Criosarasinella spinosissimum **9**,A-B
Cunningtoniceras meridionale **33**,G
Cycadolepis jenkinsiana **1**,E
Cyclothyris sp. **18**,G
Cymatoceras imbricatum **25**,A
—— *manuanense* **25**,C
Cyrtothyris undisona **46**,G

D

Damesites compactus **44**,M
Deiradoceras prerostratum **26**,E
Desmoceras latidorsatum **31**,D
Desmophyllites larteti **49**,E
Deussenia rigida **44**,J
Diadochoceras nodosocostatum **18**,D
Diaziceras tissotiforme **47**,D
Didymoceras depr. sanctaeluciensis **46**,D
—— *subtuberculatum* **48**,C
Dipoloceras cristatum **21**,F
Dosiniopsis geversi **43**,B
Douvilleiceras m. aequinodum **20**,D
Dozyia lenticularis **42**,B
Drepanochilus reineckei **28**,C
—— sp. **34**,D
Dzirulina haughtoni **20**,A

E

Elobiceras elobiense **28**,A-B
Entolium orbiculare **6**,E
Epicyprina sanctaeluciensis **20**,H
Epiphylloceras surya **51**,E
Eubaculites carinatus **49**,F, **51**,D
Eubostrychoceras indopacificum **37**,C
Eulophoceras natalense **46**,D
Eunatica sp. **22**,L
Eunerinea capelloi **29**,C
Eupsectroceras strigosum **24**,F
Euspira tealei **52**,C
Euturrilites scheuchzerianus **31**,G
Exogyra ponderosa nibelaensis **46**,C

F

Flickia quadrata **30**,C
Forbesiceras nodosum **31**,I
—— *sculptum* **31**,I
Forresteria alluaudi **38**,G
Fossarus aff. *besairiei* **22**,J

G

Gardeniceras gardeni **41**,E
Gaudryceras mite **41**,A
Gauthiericeras obesum **39**,H
Gentoniceras paucinodatum **31**,A

Gervillella dentata **16**,C
Globocardium sphaeroideum **20**,I
Glycymeris griesbachi **22**,H
—— *merenskyi* **49**,D
Goniomya umzambiensis **43**,D
Gryphaeostrea canaliculata **47**,A,**52**,I
Gymnarus auriculatus **44**,F
Gyrodes cf. *genti* **12**,C

H

Haustator scala **52**,F
Helicancyloceras crassetuberculatum **18**,F
Hemiaster forbesi **40**,D
Hemifusus umzambiensis **44**,K
Hercoglossa mazambensis **51**,C
—— *sheringomensis* **51**,A
Heteroceras elegans **13**,E
Hibolites cf. *subfusiformis* **10**,B
"Hybonoticeras" sp. **10**,D
Hyporbulites woodsi **42**,A
Hypoturrilites cricki **30**,H
—— *nodiferus* **33**,C
Hysteroceras adelei **23**,B
—— *carinatum* **23**,D

I

Idonearca? umsinenensis **20**,C
Idonearca woodsi **26**,D
Incomatiella incomatensis **52**,Q
Indogrammatodon jonesi **4**,I
Inoceramus? pedalinoides **38**,F
Inoperna baini **4**,B
Iotrigonia crassitesta **16**,A
—— *stowi* **3**,G
Isognomon ricordeanum **16**,D
—— *brentonensis* **10**,E
—— *theseni* **10**,F
Itwebeoceras lornae **39**,C

J

Jauberticeras jauberti **22**,E
Jeannoticeras jeannoti **9**,D

K

Kanabiceras septemseriatum **34**,B-C
Kangnasaurus coertzeei **2**,A
Kitchinites angolensis **48**,G
"Knemiceras" cornutum **33**,D
Kossmaticeras cf. *theobaldianum* **37**,B

L

Labeceras plasticum crassum **24**,E
Labrostrea sanctaelucialis **40**,B
—— *umsinenensis* **39**,F
Latiala bailyi **20**,F
Leiostomaster angolanus **45**,K
Leptocleidus capensis **5**,C
Lethargoceras incommodum .... **22**, B
Limnaea remota **3**,B
Linearia incomatensis **52**,R
Linotrigonia itongazi **44**,I
—— *nibelaensis* **46**,F
Littorinopsis africa **52**,O
Liopeplum capensis **44**,E
Lucina reineckei **28**,G
Lunatia multistriata **45**,G
Lycettia cf. *arrialoorensis* **52**,D
—— *uitenhagensis* **4**,C

M

Macoma? papyracea **42**,L
Mactra? dubia **7**,G
Malawisaurus dixeyi **2**,B
Manambolites dandensis **48**,A
Mantelliceras nitidum **33**,A
—— *saxbii* **30**,E
Manuaniceras manuanense **21**,B
Mariella gallieni evoluta **30**,G
Mayesella haughtoni **11**,D
—— *vau* **7**,C
Megacucullaea kraussi **4**,G
Megatrigonia conocardiiformis **10**,A
—— *obesa* **17**,G
—— *saggersoni* **15**,A

Menuites macgowani **48**,F
Meretrix uitenhagensis **7**,I
Mesocallista andersoni **20**,G
Mesoglauconia madagascariensis **22**,I
Mesopuzosia indopacifica **37**,A
Metoicoceras geslinianum **34**,G
Meyeria schwarzi **3**,D
Miodesmoceras haughtoni **7**,K
Modiolus typicus **35**,H
Monodonta hausmanni **3**,F
Morrowites mocamedensis **35**,C
Mossamedites serratocarinatus **35**,F-G
Myopholas? dominicalis **4**,K
Mytiloides striatoconcentricus **39**,A
Mytiloperna atherstoni **6**,A

N

Natalites africanus **42**,J
Natica? mirifica **3**,C
—— *uitenhagensis* **6**,G
Neithea atava **14**,D
—— *coquandi* **31**,E
—— *quinquecostata* **39**,B
—— *striatocostata* **48**,D
—— *syriaca* **19**,B
Neohibolites ewaldi **23**,F
Neohoploceras subanceps **6**,I
Neokentroceras curvicornu **28**,E-F
Neophylloceras ultimatum **50**,C
Neostlingoceras carcitanense **30**,D
Neritopsis? turbinata **4**,D
Neuquenella kensleyi **10**,H
Nilssonia tatei **1**,K
Nolaniceras uhligi **17**,A
Noramya gabrielis **14**,C
Nordenskjoeldia bailyi **39**,D
Nostoceras rotundatum **46**,A
Notoscabrotrigonia kitchini **10**,G
Nototrigonia shonei **17**,B
Nqwebasaurus thwazi **2**,C
Nuculana andradei **52**,M
—— cf. *scaphuloides* **52**,K

O

Oiophyllites angolensis **48**,G
Olcostephanus atherstonei **5**,D-E
—— *baini* **9**,C
—— *multistriatus* **9**,E
Onychiopsis mantelli **1**,I
Ophiolancea swartkopensis **5**,A
Ostlingoceras rorayense **30**,F
Otozamites rectus **1**,D

P

Pachydiscus australis **50**,A
Palaeomoera haughtoni **40**,C
Palaeozamia rubidgei **1**,J
—— *tatei* **1**,A
Paleopsephaea odonnelli **40**,I
—— *scalaris* **42**,O
Panopea gurgitis **17**,E
Paosia dutoiti **45**,B
Paranomotodom angustidens **36**,E
Paraturrilites circumtaeniatus **25**,B
Patellona kaffraria **44**,C
Peratobelus foersteri **15**,E
Perissoptera bailyi **42**,I
—— *odonnelli* **40**,E
Peroniceras dravidicum **38**,C
—— *tridorsatum* **38**,A
Pervinquieria vallifera **24**,A
Pholadomya pleuromyaeformis **18**,B
Phygraea vesicularis zululandensis **51**,B
Phyllopachyceras rogersi **6**,D
—— *whiteavesi* **31**,C
Phylloptychoceras dandense **48**,E
Pisotrigonia kraussii **4**,A
—— *salebrosa* **16**,B
Pleuromya renniei **42**,N
Plicatula andersoni **22**,K
—— *rogersi* **23**,C

Podozamites morrisi **1**,M
Poikiloceras firmum **22**,A
Pollia? gregoryi **52**,J
Praelongithyris? skoenbergensis **32**,B
—— *vanhoepeni* **25**,D
Pratulum rogersi **15**,F
Prionocycloceras carvalhoi .... **35**, A-B
Procardia vignesi **23**,J
Proeucalycoceras choffati **32**,D
Prognathodon kianda **36**,B
Prohysteroceras wordiei **28**,B
Proplacenticeras kaffrarium **38**,B
—— *umkwelanense* **37**,D
Protacanthoceras subwaterloti **32**,E
Protocardia moutai **28**,D
—— *umkwelanensis* **44**,D
Protopirula capensis **42**,M
Proveniella cf. *meyeri* **18**,C
—— *? rothpletzi* **14**,B
—— sp. **23**,I
Pseudavicula? africana **19**,E
Pseudhelicoceras catenatum **19**,F
Pseudholaster vanhoepeni **23**,H
Pseudocalycoceras angolense **34**,E-F
Pseudocucullaea lens **35**,I-J
"Pseudomelania" egitoensis **41**,F
Pseudomelania sutherlandi **45**,C
Pseudophyllites indra **45**,A
Pseudoschloenbachia griesbachi **41**,B
—— *umbulazi* **40**,F
Pseudoxybeloceras quadrinodosum **41**,C
Pseudoyaadia hennigi **11**,A
Pterophyllum africanum **1**,A
Pterotrigonia cristata **20**,B
Pterotrigonioides rogersi **7**,H
—— *savagei* **10**,C
Ptilotrigonia cricki **21**,G
—— *lauta* **19**,C
—— *lautissima* **22**,D
Ptychodus mortoni **36**,D
Ptychomya kitchini **14**,G
—— *plana* **14**,G
Puzosia australis **26**,F
—— *provincialis* **22**,G
Pygurus mendelssohni **25**,G
Pyropsis africana **43**,G

R

Rastellum deshayesi **44**,H
Reyreiceras reyrei **29**,E
Rhynchostreon matthewsi **24**,C
—— *suborbiculatum* **31**,B
Ricnoceras pandai **26**,A
Rinetrigonia laevicosta **21**,A
—— *maccarthyi* **21**,D
—— *ventricosa* **6**,F
Ringicula woodsi **43**,E
Roundairia drui **45**,F
Rutitrigonia peregrina **22**,C

S

Saghalinites cala **50**,E
Salaziceras salazense **25**,E
Sanmartinoceras africanum **13**,C
Seebachia bronni **6**,B
Sharpeiceras florancae **30**,A
Sinzovia trautscholdi **18**,A
Solariella africana **21**,C
Solenoceras binodosum **50**,B
Squalicorax pristodontus **50**,G
Steinmanella holubi **3**,A
Stoliczkaia tenuis **29**,F-G
Sphaera corrugata **14**,E
Sphenopteris fittoni **1**,F
Sphenotrigonia frommurzei **19**,G
Spiriferoid brachiopod **18**,H
Spondylus cf. *calcaratus* **44**,A
Styphloceras latidorsatum **23**,E
Submortoniceras woodsi **47**,B
Subprionocyclus massoni **35**,D-E

T

Taeniopteris sp. **1**,G
Tancredia schwarzi **4**,E

Tanzanitrigonia schwarzi **14**,A
Tegoceras camatteanum **19**,A
Temnocidaris malheiroi **29**,D
Tethyoceramus aff. ernsti **38**,D
Tetragonites subtimotheanus **24**,D
Texanites soutoni **40**,G
—— *vanhoepeni* **40**,A
Texasia cricki **42**,E
Theodoxus? choffati **52**,G
Thetis papyracea **7**,B
Tholaster carvalhoi **45**,J
Trachycardium griesbachi **42**,D
—— *reynoldsi* **44**,G
Transitrigonia herzogi **8**,A
Trigonarca capensis **43**,M
—— *elongata* **45**,H
—— cf. *ligeriensis* **31**,F
Trigonia tatei **4**,H
—— cf. *tatei* **13**,B
Trigonocallista woodsi **52**,A
Trochactaeon woodsi **45**,E
Trochoceramus andimakenensis **51**,F
Tudicla krenkeli **52**,E
"Turbo" minulutus **7**,A
Turbo rogersi **3**,H
Turrilites acutus **33**,F
—— *costatus* **30**,B

U

"Unio" uitenhagensis **1**,O
Utrobiqueostreon greylingae **11**,E

V

Vectianlla? tellina **14**,F
Velostreon umfolozianum **52**,P
Venericardia cf. *beaumonti* **52**,B
Veniella forbesiana **37**,G
Venus cordiformis **14**,E
Vepricardium? kossmati **49**,B
Vola lindiensis **14**,D
Vultogryphaea goldfussi **39**,G

Z

Zaria bonei **43**,H
Zulostrea zulu **46**,B
Zuluscaphites orycteropus **24**,G
Zulutrigonia pongolensis **15**,B